Y0-BBX-738

John L. Cloudsley-Thompson

Evolution and Adaptation of Terrestrial Arthropods

With 86 Figures

Springer-Verlag
Berlin Heidelberg New York
London Paris Tokyo

Professor Dr. JOHN L. CLOUDSLEY-THOMPSON
Department of Biology (Medawar Building)
University College
University of London
Gower Street
London WC1E 6BT, GBR

Cover illustration:
European woodlice: *Trichoniscus pusillus; Philoscia muscorum; Oniscus asellus; Porcellio scaber; Armadillidium vulgare.* This volume, page 31, Fig. 15 a–e.

ISBN 3-540-18188-1 Springer-Verlag Berlin Heidelberg New York
ISBN 0-387-18188-1 Springer-Verlag New York Berlin Heidelberg

Library of Congress Cataloging-in-Publication Data. Cloudsley-Thompson, J. L. Evolution and adaptation of terrestrial arthropods / John L. Cloudsley-Thompson. p. cm. Includes index. ISBN 0-387-18188-1 (U.S.) 1. Arthropoda–Evolution. 2. Adaptation (Biology) I. Title. QL434.35.C55 1987 595.2'0438–dc 19   87-28666 CIP

© Springer-Verlag Berlin Heidelberg 1988
Printed in the United States of America.

2131/3130-543210

Preface

This is not intended to be a comprehensive text book of entomology and arachnology, but rather a concise synthesis of certain basic information required for BSc (Hons) and MSc (Entomology) examinations. The approach is primarily functional: for instance, the skeletal and waterproofing properties of the arthropod cuticle are discussed, but not its biochemistry; and I have included only those points with which I believe all advanced students of the subject ought to be familiar. Some aspects are today regarded as outdated; others do not appear in any current texts, but I have included them because I consider them to be important. In no way, therefore, should this be regarded as a book of reference. To be frank, it consists of a mass of oversimplifications and unqualified generalizations which are intended to clarify the complex principles underlying them. Once these principles have been thoroughly grasped, the reader will have acquired a sufficiently broad approach to be able to get the best value from more advanced treatises.

My thanks are due to Drs. John Dalingwater, Andrew Milner, and especially Paul Hillyard for their advice on palaeontological matters; to Professors Einar Bursell for permission to use Fig. 8, taken from his *An Introduction to Insect Physiology* (Academic Press), Neil F. Hadley and the Editors of *American Scientist* to reproduce Figs. 52, 53, and to F. Schaller for Fig. 42 from a chapter he wrote in Gupta (ed.) *Arthropod Phylogeny* (Van Nostrand Reinhold). Finally, I would like to express my gratitude to Roy Abrahams for redrawing the figures showing insect wing venations and to Eileen Bergh for typing the manuscript. The book was completed during tenure of a Leverhulme Emeritus Fellowship which I am pleased to acknowledge, while Professor N.A. Mitchison FRS kindly provided me with accommodation in University College, London.

J.L. CLOUDSLEY-THOMPSON

Contents

1 Palaeontology and Phylogeny

Millions of years before arthropods had succeeded in colonizing the land, the primaeval oceans were teeming with them. They, in turn, had probably evolved from segmented, worm-like ancestors, similar to today's polychaete annelids; but fossil evidence of this is lacking. Trilobites and Crustacea are known from the Cambrian period, Eurypterida from the Ordovician, Diplopoda from the Devonian or possibly earlier, Collembola and probably Insecta also from the Devonian. Myriapods, Collembola, and Insecta are probably descended from an onychophoran-like ancestor. The mid-Cambrian *Aysheaia pedunculata* (found in British Columbia in a marine deposit, along with polychaetes, trilobites, holothurians etc.) may have been such an animal, but other interpretations of it have also been made.

1.1 The Earliest Arthropodan Fossils

The fossil record of the Arthropoda extends back to the Early Cambrian. Even at that time, however, the main taxonomic groups were already in existence as separate entities. Within these main groups were numbers of subgroupings, some of which are still in existence, others now extinct. The earliest divergence of the arthropods at a high level must, therefore, have taken place before the first known specimens were fossilized. The Chelicerata, Uniramia, Crustacea and Trilobita arose from amongst a plethora of arthropod types, mostly now extinct. We must probably look back to Pre-Cambrian soft-bodied forms for the common ancestors of these major groups.

The similarity between the head of *Aysheaia* and the prostomium of existing annelids suggests that the Onychophora may have originated from lobopod annelids which had already evolved a prostomium. The transition from these must have taken place in the oceans of the Early Palaeozoic. It involved loss of the dorsal or notopodial plates bearing chaetae, the loss of segmentation of the integument and musculature, and the disappearance of an eversible proboscis. Jaws evolved from the lobopodia of the third segment, those of the fourth being reduced to small oval papillae. The lobopodia of the fifth and subsequent segments became legs that differed little from annelid parapodia. The appearance of gill outgrowths in the lobopodia of marine Onychophora was undoubtedly the result of sclerotization of the integument, which led to considerable loss in respiratory function. When the Onychophora became terrestrial, they lost all these gills except for their most basal outgrowths, which were drawn into the bases of the legs and became eversible sacs.

Whether or not the Crustacea (which use the gnathobases of their limbs for biting) should be included with the other mandibulates, is disputable since, in Onychophora

(*Peripatus*), myriapods, and insects, a whole limb has become the jaw and the tip is used for biting (see below).

The origin of the chelicerates is even more problematical. Despite their diversity, they possess a large number of characters in common. From the oldest of the merostomes, it is easy to derive the scorpions. These resemble Eurypterida in the number and distribution of their segments and tagmata and, also, in possessing a narrow postabdomen. Nearly all recent attempts at phylogenetic analysis of the Chelicerata indicate that the Eurypterida are the sister-group to either Scorpionida or to all Arachnida, but not to Xiphosura alone. Little is known of the relationships of the Pycnogonida. Some zoologists consider them to be a class of chelicerates; others regard them as a separate subphylum of the Arthropoda. An older view, relating the sea spiders to the Crustacea has been abandoned, since nauplius and proto-nymphon larvae are dissimilar. In the absence of a fossil record, the problems of pycnogonid origins cannot be solved satisfactorily, and they do not appear to be closely related to any other group of arthropods.

Myriapods (Archipolypoda) appeared in the Silurian and Devonian periods, Diplopoda and Chilopoda are represented in the Devonian. The first terrestrial arachnids were early Devonian; but aquatic groups, like Eurypterida and Xiphosura, are much older, playing a prominent role in the Early Cambrian. Scorpions appeared in late Silurian and Devonian times, and are similar to eurypterids. Other arachnids cannot be traced with certainty to any aquatic ancestry nor, for that matter, can the insects.

1.2 The First Terrestrial Arthropods

The earliest records of terrestrial animal life occur in Silurian deposits, but the record is sparse. Silurian and Devonian fossil scorpions, long believed to be the earliest land animals, are now known to have been fully aquatic, and to have shared the marine or brackish sublittoral zone with their close relatives, the eurypterids. At the same time, some of these and a few xiphosurans had adaptations suggesting that they were capable of brief excursions on land.

Myriapods (Archipolypoda) from the Lower Old Red Sandstone of Britain appear among the earliest terrestrial fossils, for the identity of the Silurian forms is uncertain. It has long been thought that *Archidesmus loganensis* and *Necrogammarus salweyi*, for example, were myriapods. The former has the appearance of a juliform millipede, but it also resembles some of the plant debris found in the same geological horizon. *N. salweyi* (Fig. 1) was first thought to be an amphipod crustacean, and later referred to the Archipolypoda. It has recently been shown to be the infracapitulum (fused labrum and palpal coxae) and palp of a large pterygotid eurypterid. There is a considerable amount of fossil myriapod material from the Devonian Old Red Sandstone, including flat-backed, polydesmid-like types such as *Archidesmus macnicoli*, and rounder-bodied juliform types including *Kampecaris forfarensis*.

Vascular land plants first appeared in the Silurian Period (there may even have been some in the Ordovician) and herbivores would have emerged from the seas almost immediately afterwards – among them were probably millipedes, Collembola, Thysanura and mites. Carnivores and predators, comprising third and fourth trophic-level

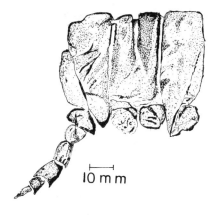

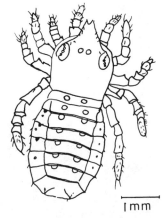

Fig. 1. *Necrogammarus.* (After Rolfe 1980)   **Fig. 2.** *Palaeocharinus.* (After Rolfe 1980)

consumers followed. The arthropods of the Rhynie Chert, near Aberdeen, one of the earliest land fossil beds, included springtails (*Rhyniella*), mites (*Protacarus*), spiders (*Palaeocteniza*) and the long-legged trigonotarbid arachnids *Palaeocharinus* and *Palaeocharinoides* (Fig. 2). The collembolans and mites might have fed on micro-organisms and spores in the soil, while the other arachnids may have been carnivorous. On the other hand, since their fossils have been found within hollow stems and sporangia, it is equally possible that they, too, were originally spore-feeders which conserved their body moisture by inhabiting damp rotting crevices. Later they turned their attention to other micro-arthropods seeking shelter in these humid niches, and eventually adopted a predatory existence. They might have lurked in dehisced sporangia, while these were still borne upright upon the parent plants, and here awaited the arrival of food in the form of palaeoarthropod spore-feeders.

The Lower Devonian fauna of Alken an der Mosel, in Germany, a somewhat later fossil bed, contained a mixture of aquatic, amphibious and terrestrial animals that lived in or beside a land-locked lagoon communicating with the sea at high tide. Many of these Eurypterida were probably able to leave the water temporarily and move across the swampy land. Even those, such as *Parahughmilleria,* which were well adapted for swimming, may have used their paddles for crawling on land, as one of their fossilized trails suggests. The eurypterids were huge predators, apparently lords of the shallow seas. Their victums included the earliest vertebrates, the Ostracodermi, which probably gained some protection from the bony plates that covered their bodies. When these were replaced by more sophisticated swimmers, some of the Eurypterida moved into fresh water or became terrestrial. Certain genera lingered into the Permian. It is noteworthy that terrestrial arachnids, such as *Alkenia mirabilis* (order Trigonotarbi) existed in the Lower Devonian at this time, while the forerunners of the scorpions, such as *Waeringoscorpio*, were still aquatic and breathed by means of gills. This suggests that different contemporary arachnids must have at least two ancestral origins.

Genuine myriapods appeared first in the Scottish Old Red Sandstone, the relevant part of which is believed to be Silurian rather than Devonian, as was presumed to be

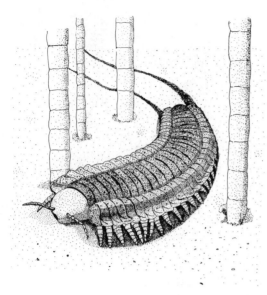

Fig. 3. Reconstruction of *Arthropleura* making its trail. (After Briggs et al. 1984)

the case until recently. Rich deposits of fossil plants and animals of Devonian age have been found near Gilboa, New York, which, although younger than the Rhynie Chert and Alken beds, is richer in taxa. Fragmentary remains and nearly whole specimens have been assigned to Eurypterida, Arachnida (Trigonotarbida, Pseudoscorpiones and Acari: the presence of spiders and sorpions is equivocal). Chilopoda (possibly Craterostigmatomorpha and Scutigeromorpha) and arthropods that could perhaps be Insecta (Archaeognatha) have also been found: These are probably the earliest records for these groups. During the Carboniferous, a giant myriapod, *Arthropleura,* left tracks in the mud as it weaved through forests of *Calamites* (Fig. 3).

1.3 Evolution in the Arthropods

1.3.1 Embryological Considerations

The fossil record tells something about the existence of certain extinct arthropod groups, but very little about their inter-relationships. Thus, although the phylogeny of some of the chelicerate groups is not entirely obscure, palaeontology reveals almost nothing about their relationships with the Crustacea and the terrestrial Uniramia — Onychophora, myriapods and hexapods. Comparative morphology and embryology, on the other hand, do give some evidence on this subject, although even this cannot be interpreted unequivocally.

  Information derived from study of the comparative functional morphology of limbs, jaws, and the organization of the anterior end of the body, demonstrates basic similarities between each of the three taxa, Chelicerata, Crustacea and Uniramia. The same evidence, however, also shows how great are the differences that separate them — differences so great that S.M. Manton and others have considered that the rank

of phylum should be applied to each. The clitellate annelids (Oligochaeta and Hiru-
dinea) display a form of development in which spiral cleavage has been retained,
although the trochophore larva, characteristic of the Polychaeta, has been eliminated.
Development is direct, into a metamerically-segmented body. The presence of a clitel-
lum in annelids has enabled the intralecithal cleavage and blastoderm development
characteristic of the Onychophora to evolve. The yolk mass remains undivided while
the zygote nucleus and its daughters divide and spread, with accompanying divisions
of their cytoplasmic haloes.

The onychophoran pattern of development underlies all the diversity and specialization exhibited within the Myriapoda and Hexapoda — supporting the concept of
their phylogenetic unity — but it is clear not only that the Onychophora are not
ancestral to the other two groups but, rather, that each has diverged independently
from a common lobopod ancestry.

The Crustacea are distinct from the myriapods and insects except in that their
embryological development takes place by spiral cleavage. The third major taxon of
Arthropoda is the Chelicerata, whose specialized total cleavage and configuration of
presumptive areas obscure any evidence of past relationships. The chelicerates are
undoubtedly not closely related to the Uniramia, but links with the Crustacea or, for
that matter, with the Annelida, cannot be established embryologically.

1.3.2 Comparative Morphology

Two different types of evolutionary advance can be recognized within any particular
taxonomic group of animals. One consists of morphological changes associated with
the assumption of a different habit or mode of life: the other of adaptations to a
particular habitat. The first of these gives rise to major phyletic changes that are often
regarded as evolutionary 'advances'; the second leads to adaptive radiations within a
taxon. For instance, the development of a plastron enables certain Hemiptera to live
in water and is one of their adaptations to an aquatic habitat. It is probable that
adaptive radiations of this kind have had little or no effect on the evolution of larger
taxa but, instead, represent the recently diversified branching of successful groups of
animals.

Evolutionary advances concern the assumption of habits which do not restrict
their possessors to any particular environment. They enable animals to survive better
in a variety of habitats. For example, beneath a single rock or rotting log may be
found a diverse array of woodlice, springtails, beetles, mites, false-scorpions, centipedes and millipedes. Although these animals are living in the same environment,
they can survive in many other habitats as well. They are morphologically distinct
from one another, and are not all restricted to the same environment as they would
be if they represented adaptive radiations within the same arthropod group.

The feeding mechanisms of chelicerates differ from those of other arthropods. In
Xiphosura, paired gnathobases on the five prosomal limbs behind the mouth and
sometimes the coxae of the first legs are used, in the Arachnida only those on the
pedipalps. Apart from the location of the gnathobase on the coxa, there is nothing
really in common in this respect between the Chelicerata and the Crustacea. Leg

morphology, gaits and limb-base mechanisms have been studied in great detail, and emphasize basic distinctions between the Uniramia, Crustacea and Chelicerata. In the Uniramia there are whole-limb jaws and basal lobes are not used for biting; only the tips of the jaws. This whole-limb mandible is unjointed in the hexapods, and the primitive movement is rolling, grinding – derived from the promotor-remotor swing of a walking leg. In the myriapods, on the other hand, the jaws bite transversely. Thus the three types of jaw in the Chelicerata, Crustacea, and Uniramia suggest probable independence in the evolution of these three groups, and no one can have given rise to another. A further subdivision separates the hexapods and the myriapods from one another, and from the Onychophora whose whole-limb mandible slices the food by an antero-posterior movement similar to the stepping of the locomotory legs. There is, however, convergence between crustaceans and hexapods (both of which basically have rolling jaws) to secondary transverse biting jaws, and between hexapods and Myriapoda to ectognathous protrusible jaws.

1.3.3 Monophyletic or Polyphyletic Origins

Whereas S.M. Manton (1964, 1977) regarded the arthropods as being polyphyletic for the reasons mentioned above, many other biologists consider them to be mono-phyletic. Indeed, most cases of alleged polyphyly reported in the literature do not stand up under critical analysis. The point at issue is whether the boundary between ancestral and descendent groups has been crossed more than once. Since that bound-ary is arbitrarily designated by the choice of one or more diagnostic characters, groups which are polyphyletic in one sense can always be made monophyletic by choosing different, more widely distributed diagnostic characters. At the same time, arthropod polyphyly does not necessarily imply independent evolution of all arthropod charac-ters, and Manton showed that the main locomotory and feeding mechanisms of the four main arthropod phyla were derived independently from different soft-bodied ancestors. These may well have had a common ancestor themselves, perhaps even with compound eyes, but the fact that they had soft bodies and were not arthropodized means that they, themselves, were not arthropods and, therefore, the later arthropod grade is polyphyletic. The Middle Cambrian Burgess Shale shows many odd arthropods with a mosaic of characters, as well as some soft-bodied arthropod-like organisms (e.g. *Opabinia*).

A polyphyletic origin, in the sense first defined, would imply that all important arthropodan characters had evolved many times independently – the head composed of six segments, compound eyes and other sense organs, mouth parts, legs and so on. One of these basic characters is a photoreceptive system composed of median and lateral, faceted, compound eyes. These sense organs can be found in all major recent and fossil arthropodan taxa but, to establish their monophyletic relationships, it is necessary to compare them in detail with special reference to similarities which have not been acquired through parallel evolution (Fig. 4).

The Xiphosura (e.g. *Limulus*) are the only Recent chelicerates with lateral faceted eyes, but all terrestrial Arachnida have lateral ocelli – up to five on each side of the prosoma. The eyes of scorpions can be derived from the type of eye found in *Limulus*

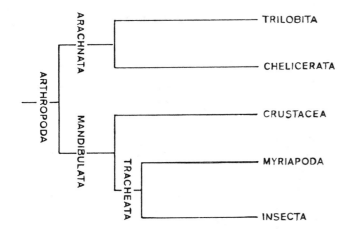

Fig. 4. Dendrogram showing phylogenetic relationships of various arthropod groups. (Based on Paulus 1979)

by diminution and separation of the complex into five parts, and fusion of the remaining ommatidial facets into five big lenses. Numerous isolated ommatidial rhabdoms can be distinguished beneath these lenses and in other Arachnida can be found a further modification of the situation that obtains in the Scorpiones. This conclusion is supported by the similarities in the embryonic development of all arthropodan eyes. The ommatidium probably evolved only once within the Crustacea and Insecta because it consists in each case of a cornea composed of two corneagenous cells, a crystalline cone and a retina of eight cells. Myriapod eyes, like the eyes of larval insects, are modified ommatidia. An ultra-structural study of the compound eye of *Scutigera* (Fig. 5) has shown that this is secondarily faceted, and therefore not homologous with other compound eyes, but this has been disputed.

The ommatidium has therefore evolved only once. The same is true of median eyes and frontal organs, which are reduced in myriapods. This indicates that the Mandibulata, at least, are monophyletic. The relationship between the Mandibulata and the Chelicerata is less clear — the eyes of *Limulus* are constructed too simply to provide enough characters for a detailed comparison. Nevertheless, it seems probable that the Chelicerata and Mandibulata may also be monophyletic because both groups have a basic number of four median eyes, but it cannot be proved with certainty from a comparison of the lateral ommatidia.

Further evidence for a monophyletic origin of the Arthropoda is afforded by the ultrastructure of sperm. Basic aquatic sperm is present only in Tardigrada and Xiphosura. (The Onychophora possess highly evolved spermatozoa and, therefore, appear as a sister group to the arthropods.) The spermatozoa of the lowest Arachnida, Crustacea and Myriapoda appear to have evolved only slightly from the basic type. Similar conclusions have also been reached from embryological studies and from considerations of comparative morphology and physiology including the intersegmental tendon system, neuroendocrine structures, and haemocytes. These will not be discussed here, as detailed evidence is presented in the book edited by Gupta (1979). It is, however, worth mentioning that the primitive visceral pattern found in some arthropod groups not only unites the Chelicerata with the Crustacea and Uniramia, but also links them

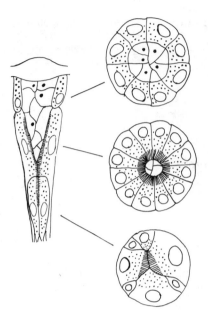

Fig. 5. Ommatidium of the compound eye of *Scutigera*.
(Modified after Paulus 1979)

to the Tardigrada, Pentastomida and Pycnogonida. Other features that unite all these arthropods include: (i) the haemocoele with horizontal pericardial septum and tubular osteate heart, (ii) the coelomic organs, (iii) the paired, sac-like gonads, (iv) the ectodermal foregut and hindgut with their unusual muscular coats contrasting with those of the midgut. In addition, they are connected by the absence of flame cells or nephridia, and the lack of specialized organs for nitrogenous excretion, and salt and water balance. K.U. Clarke (1973) concluded that the primitive visceral pattern found in some arthropod groups links this phylum with the Aschelminthes, and that it is among these animals that lines leading to full arthropod design must have originated.

From the considerations discussed above, it will be understood that the phylogeny of the arthropods remains a vexed question through lack of adequate palaeontological data. The balance of evidence at present available, including that from palaeontology, embryology and comparative morphology, does not, however, justify classifying the three taxa Uniramia, Crustacea and Chelicerata, as three separate phyla: they are probably best regarded as subphyla of the phylum Arthropoda which is distinct from the phylum Onychophora. The simplest theory of arthropod origins is that the group is monophyletic, but characterized by features, such as arthropodization, which may well have arisen in parallel.

Further Reading

Almond JE (1985) The Silurian – Devonian fossil record of the Myriapoda. Phil Trans R Soc (B) 309:227–237

Anderson DT (1973) Embryology and phylogeny of Annelids and Arthropods. Pergamon Press, Oxford

Briggs DEG, Plint AG, Pickerill RK (1984) Arthropleura trails from the Westphalian of Eastern Canada. Palaeontology 24:843–855

Clarke KU (1973) The biology of the Arthropoda. Edward Arnold, London

Gupta AP (ed) (1979) Arthropod phylogeny. Van Nostrand Rheinhold, New York

Kevan PG, Chaloner WG, Saville DBO (1975) Interrelationships of early terrestrial arthropods and plants. Palaeontology 18:391–417

Manton SM (1964) Mandibular mechanisms and the evolution of the arthropods. Phil Trans R Soc (B) 247:1–83

Manton SM (1977) The Arthropoda. Habits, functional morphology and evolution. Clarendon Press, Oxford

Paulus HF (1979) Eye structure and monophyly in the Arthropoda. In: Gupta AP (ed) Arthropod phylogeny. Van Nostrand Reinhold, New York, pp 299–383

Wolfe WDI (1980) Early invertebrate terrestrial faunas. In: Panchen AL (ed) The terrestrial environment and the origin of land vertebrates. Academic Press, London New York (Systematics Association Special Volume No. 15, pp 117–157)

Rolfe WDI (1982) Ancient air breathers. Field Mus Nat Hist Bull 53:12–16

Shear WA, Seldon PA (1986) Phylogenetic relationships of the Trigonotarbida, an extinct order of arachnids. Actas Congr Int Arachnol Jaca España 1986, 1:393–397

Shear WA, Bonamo PM, Grierson JD, Rolfe WDI, Smith EL, Norton RA (1984) Early land animals in North America: evidence from Devonian age arthropods from Gilboa, New York. Science 224:492–494

Whittington HB (1979) Early arthropods, their appendages and relationships. In: House MR (ed) The origin of major invertebrate groups. Academic Press, London New York (Systematics Association Special Volume No. 12, pp 253–268)

2   Implications of Life on Land

2.1 The Significance of Size

Depending upon their sizes, animals are affected in different ways by the various physical factors of their environments. Molecular forces, Brownian movement and the viscosity of water are extremely important to small protozoans, while gravity becomes increasingly important in the energetics of larger animals. The results of natural selection must be quite different on smaller than on larger organisms.

In a comparison between social human beings and social ants, the point has been made that the difference in size between the two has unexpected consequences. For instance, ants could not use fire because even the smallest fire that is stable must be larger than an insect. Nor could ants keep a wood fire burning, because they are too small to get near enough to add fuel, even if they were able to carry it. Again, ants cannot use tools. A miniature hammer would have too little kinetic energy to drive even a miniature nail, while ant-sized spears, clubs and arrows, would be equally ineffective. Ant-sized books would be impossible to open because their thin pages would be held together by intermolecular forces that are relatively extremely powerful at that scale. In any event, reading would not be possible for ants because they do not have enough brain cells. Finally, ants cannot wash themselves with water because the surface tension would interfere and, in any case, the insect integument repels water. When they do inadvertently get into water, ants cannot easily extricate themselves. Instead of washing, they dry-clean themselves by rubbing dry particles over their bodies and then scraping them off again!

Nevertheless, many small insects, such as pond-skaters (Gerridae), are able to walk and run on the surface of water. They are supported by surface tension acting on their feet. Figure 6 shows how the foot rests on the water, and how the surface is deformed by it. The contact angle (α) of water with the cuticle is high so that, if the foot is pushed well down into the water, surface tension (T) acts more or less vertically upon

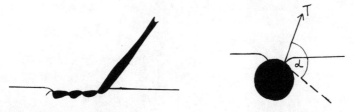

Fig. 6. Diagram showing the foot of an insect standing on the surface of water. *Left* in side view; *right* in section. (After Alexander 1971)

it. Whereas this is quite sufficient to support a relatively small insect, a human being, weighing more than a million times more, would need feet with a total perimeter of over 7 km. This is because the force exerted by surface tension is proportional to the length of the feet, while the weight that has to be supported is, in isometric animals, proportional to the cube of that length (see Sect. 2.1.1). The fact that the integument is hydrophobic may, perhaps, have a bearing upon the origin of wings from the tracheal gills of aquatic insects (Chap. 4.4).

2.1.1 Size, Skeletons and Allometry

Changes of size make necessary changes in the structure and in the proportions of various parts of the body of an animal. If two organisms were of exactly the same shape, so that one was a precise scale model of the other, they would be *isometric* in all respects. Every linear measurement of the larger would be n times that of the smaller − the lengths and diameters of the body, limbs, internal organs and so on. However, the surface areas of all these parts, like that of the whole body of the larger organism, would be n^2 times that of the smaller; while its weight would be n^3. Conversely, the lengths and breadths of corresponding parts of isometric organisms would be proportional to (body weight) $^{0.33}$, areas to (body weight) $^{0.67}$ and volumes or weights equal to (body weight) $^{1.00}$. These simple relationships are responsible for the differences in shapes, structure and physiology between larger and smaller animals of the same groups or taxa.

The two main types of skeleton found in the animal kingdom are directly related to the sizes of their possessors. These are the exoskeletons of Arthropoda and the endoskeletons of vertebrate animals. An exoskeleton, which is basically a tubular structure, is probably more efficient for smaller animals; solid bones for larger forms. Tubes are extremely resistant to twisting and bending − a straw can withstand a surprising strain before it snaps − but, above a certain size, become disproportionately heavy. Tubular structures are excellent for scaffolding, but great weights have to be supported by solid girders.

Among arthropods, only marine forms, whose bodies are supported by the water, are able to attain any great size or weight. The giant Japanese spider crab *Macrocheira kaempferi* has a leg span of up to 3.8 m and may weight 6.35 kg; but the largest terrestrial scorpion, *Pandinus imperator,* only reaches a length of 24 cm and a weight of 50 g. The heaviest spider ever recorded, *Theraphosa leblondi,* weighed less than 85 g! In contrast, the largest extinct scorpion, *Gigantoscorpius willsi,* which lived in the sea of the Silurian period, reached a length of 36 cm while the largest Silurian eurypterid was 2 m long. There is comparatively little overlap in size between animals with exoskeletons and those with endoskeletons. (Even the smallest of the mammals, such as Savi's pygmy shrew, which has a length of 3.8 cm and weighs some 2.5 g and the smallest bat, a West African pipistrelle weighing about the same, are both larger than the majority of arthropods.)

The largest living insect is probably the heavily-armoured *Goliathus regius* of Equatorial Africa, adult males of which measure up to 11 cm in length and nearly as much in width. Their weight ranges between 70 and 100 g, while the elephant beetle, *Megasoma elephas* of Central America, measures up to 1.3 cm in length but weighs less because more than a quarter of its length is taken up by the cephalic horn. At the other end of the scale, some parasitic Hymenoptera (Mymaridae), hairy-winged beetles of the family Ptilidae (= Trichopterygoidea) and certain mites are considerably less than 0.25 mm in length, despite the complexity of their structure. Some are so small that they weigh less than the nucleus of a single large protozoan cell. On the whole, the lower limits of size in different animal taxa are more constant than the upper, and that of the vertebrates is far above that of any other group. The sclerotised exoskeleton and its concomitants in relation to size are by far the most important factors affecting the physiology and adaptations of terrestrial arthropods.

The need for ecdysis limits absolute size because, in larger animals, a considerable increase in weight results only in a relatively small gain in linear dimensions (Sect. 2.3) and the integument becomes disproportionately heavy. This is especially important in the case of terrestrial arthropods. At the same time, moulting is a hazardous event at all times. If all does not go exactly right, the animal may die; and it is always extremely vulnerable to its enemies until its new cuticle has hardened. So, the benefits conferred by an exoskeleton are increasingly offset by its disadvantages when the size of its possessor increases. It is highly improbable that any animal reaches the physical limitations inherent in its structural design, as it would lose its competitive efficiency before then. Selection operates not on single parameters, but upon combinations of sometimes incompatible factors.

2.1.2 Allometric Growth

Like many other animals, insects exhibit the phenomenon of allometry. When they increase in size, certain parts of the body become disproportionately large. Thus, the workers of many species of ants develop enormous heads and mandibles. In driver ants (*Anomma*), the largest workers may be 75 times larger than the smallest. The males of exotic rhinoceros beetles often bear on their thorax antler-like processes which, in some cases, are so large that they appear to be a hindrance to their owners. So, too, the mandibles of the dobson fly, *Corydalis cornuta,* are exceedingly long and cruciate. Changes in size always make necessary some changes in shape and proportions. In extreme cases, the linear dimensions of certain parts change, during growth, proportionately very much more than do the dimensions of the whole body. The significance of size and proportions is outlined in simple terms by Alexander (1971).

2.2 Water Relations

On account of their relatively small body sizes, terrestrial arthropods have extremely large surface/volume ratios. This imposes considerable problems for life on land, because the surface through which water vapour can be lost by evaporation to the environment is very large in relation to the water reserves which must sustain that loss.

There are two obvious ways in which small animals can evade desiccation on dry land. One is to avoid dry places and remain most, if not all, of the time in a humid

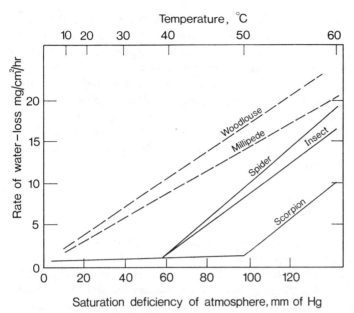

Fig. 7. Rates of water loss in dry air at different temperatures and corresponding saturation deficiencies from a woodlouse, millipede, spider, insect and a scorpion. In the woodlouse and millipede the rate of water loss is proportional to the saturation deficit of the atmosphere; but in the spider, insect and scorpion, it is negligible below a critical temperature, at which their epicuticular wax layers became porous. (Cloudsley-Thompson 1977)

environment; such a strategy has been adopted by the Crustacea. The other is to evolve an impervious integument. Both methods have been exploited by the Arthropoda in the conquest of the land, and each has its advantages and drawbacks. Indeed, on the basis of this character, the terrestrial members of the phylum can be divided roughly into two main ecological groups: the first includes woodlice, centipedes, millipedes and their allies, which lose water comparatively rapidly in dry air; the second, the arachnids and insects which are covered with an epicuticular layer of wax that reduces transpiration and renders them comparatively independent of most surroundings (Fig. 7).

Forms lacking an epicuticular wax layer are almost entirely nocturnal in habit and, in general, can wander abroad only after nightfall when the temperature drops and the relative humidity of the atmosphere rises. In contrast, most insects, spiders and other arachnids are potentially day-active except, perhaps, in deserts and other regions with rigorous climates, where the majority of invertebrates avoid the excessive midday heat and drought by their nocturnal behaviour. It seems probable that existing primitive taxa such as cockroaches, Dermaptera, Embioptera, scorpions, whip-scorpions, and spiders of the families Liphistiidae, Theraphosidae, Dysderidae and so on, have probably become secondarily adapted to nocturnal habits as a result of competition with more advanced forms. At the same time, however, many of them are large and somewhat vulnerable to enemies so that they may need to escape the attentions of potential predators in this way.

It would be mistaken to regard the absence of a cuticular wax layer as being a primitive characteristic, although the forms that lack one are so restricted in their choice of environment that they cannot be regarded as entirely successful land animals. Rather, it seems that different methods have been exploited for surviving the conditions of life on land. Animals crossing the sea-shore may well be subjected to high temperatures, and the ability to lose heat by evaporation of water may have had considerable survival value. These ancestral crustaceans were probably at first restricted to damp environments by their behaviour mechanisms. Later, their descendants exploited still further this form of terrestrial life while the insects and arachnids acquired waterproof integuments.

2.3 The Conquest of the Land

The ancestral Crustacea probably came on land across the littoral zone in comparatively recent times. Shore-dwelling species are still the most primitive in structure. There is, however, little evidence to suggest that the ancestors of insects and chelicerates did the same. Indeed, it seems more probable that primaeval scorpions and myriapods crawled onto the mud surrounding tropical swamps (Chap. 1.2), while many of the earliest forms of other arthropodan taxa may have inhabited the damp soils beneath plants growing in the Silurian period. They would, in many ways, have paralleled the cryptozoic fauna of the leaf litter and soil of forests today, living sheltered lives from which the rigours of a true terrestrial existence were largely excluded.

This suggestion implies that the ancestors of modern terrestrial arthropods invaded the land on several occasions and in more ways than one. There is no difficulty in accepting this hypothesis, since it is supported by the palaeontological evidence presented in Chapter 1. All existing arachnids are terrestrial. The existence of scorpions as fossils in earlier rocks than other terrestrial arachnids does not, however, necessarily imply that they were the first terrestrial arachnids. Almost equally primitive characters are to be seen in Schizomida, Palpigrade, Uropygi, Amblypygi and Solifugae; but these are rare in the fossil record, and generally appear much later. Perhaps their fossilized ancestors have not been found either because they were very small, or because they inhabited environments where fossilization was unlikely. Yet several of the arachnids of the Rhynie Chert were extremely small (Chap. 1.2), so this may not be the answer.

2.4 The Integument

The arthropod cuticle functions both as a body covering and as an exoskeleton. Consequently the material of which it is composed must embody a number of diverse properties. It must be rigid, except at points of articulation where it needs to be flexible and, to be an effective integument for terrestrial forms, it must be impervious to water. These varying requirements are served by the different layers of which it is composed (Fig. 8), each of which has its own characteristics.

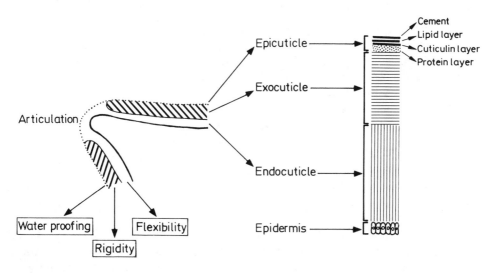

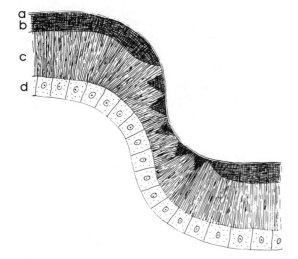

Fig. 8. The insect cuticle: a diagrammatic summary of its functions. (*Above*
Bursell 1970). *Below* note the reduction
of the exocuticle where articulation
takes place and where rigidity is not
required: here it is broken up into
wedge-shaped pieces. *a* waterproofing
epicuticle; *b* resistant exocuticle; *c* flexible endocuticle; *d* epidermis

2.4.1 The Endocuticle

This is a lamellated structure composed principally of protein microfibrils and the
polysaccharide chitin. The microfibrils are arranged in layered sheets, each oriented in
a different tangental plane. Only endocuticle and epicuticle may be present in regions
where articulation takes place, or else the exocuticle is reduced (Fig. 8).

2.4.2 The Exocuticle

This is really homologous with an outer layer of endocuticle which has been hardened
and partially waterproofed by the process of sclerotization. For this reason, the two

layers are regarded as being components of a single layer, the 'procuticle' which is distinct from the epicuticle. In the process of sclerotization, the protein constituents of the exocuticle are tanned by quinones to form 'sclerotin'. This can be seen to take place in a newly-moulted insect where it gradually hardens, and turns brown as a result of melaniza-tion. The process is analagous to that which takes place when raw hide is converted into leather by soaking it in a solution containing tannic acid; hence it is often spoken of as 'tanning'. The resulting sclerotin is tough, resistant, and light in weight — essential qualities in the exoskeleton of a terrestrial animal. In addition to being sclerotized, the cuticles of crustaceans are hardened by the deposition of calcium carbonate, and the same is true of Diplopoda. The extra weight is an advantage to marine Crustacea which would otherwise be too buoyant and might easily be swept away; whilst the extra hard-ness afforded to millipedes is an adaptation to pushing through leaf litter and under bark. Millipedes have plenty of legs on which to distribute their excess weight!

Various modifications of the integument allow for flexibility and extension, as already mentioned. For example, the hardened exocuticle may be broken up into wedge-shaped pieces at intersegmental membranes to permit bending (Fig. 8) or may be absent in bugs and other insects that take large meals at irregular intervals. In these, the epicuticle is much folded to allow for the expansion of the abdomen during feed-ing. The hyaline exocuticle of scorpions may be responsible for their fossilization. In beetles, the hardness of the integument is often enhanced by the presence of a network of strands known as 'balken' embedded in a homogenous matrix of the endo-cuticle. Certain ticks, Thysanura, Orthoptera (Chap. 6.1.2), flea prepupae, Psocoptera, Siphonaptera and other insects are able to absorb moisture from the atmosphere, even when the air is unsaturated, and without losing their resistance to desiccation. This subject has been discussed in detail by Edney (1977).

2.4.3 The Epicuticle

This, the outermost layer of the integument, comprises no more than 5% of the total thickness. It differs from the layers of the procuticle in that it does not contain chitin; but it is, in fact, much more complex in structure. Four distinct layers can usually be distinguished by histological techniques. These are as follows: (a) An outermost layer of resin-like 'cement' which protects the layer beneath. (b) A layer of lipid, the 'wax layer'. (c) A 'cuticulin' layer which forms a substrate for the lipid. (d) A 'protein layer' composed of sclerotized proteins impregnated with lipids but containing no chitin. The wax layer is the most significant water-proofing component of the cuticle, and it is protected from abrasion by the overlying cement layer.

An interesting example of this occurred in 1969 when C.S. Crawford and I were working at the University of New Mexico on the 'vinegaroon' *Mastigoproctus giganteus*. Much to our surprise we found that these large 'pedipalpi', although they inhabited the Chihuahuan desert, died rapidly from desiccation in dry air, and experiments on transpiration at different temperatures failed to disclose the critical point (Fig. 9). These results conflicted with those of G.A. Ahearn, who, unknown to us, was working on the same species at Arizona State University, Tempe, but obtained a typical water loss curve with a critical point around 37.5°C. The apparent paradox was explained by

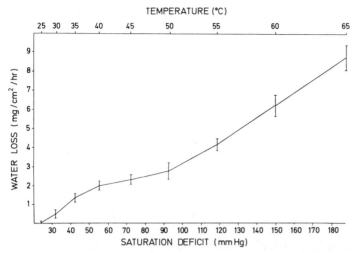

Fig. 9. Transpiration from *Mastigoproctus giganteus* in relation to temperature of the ambient air. Each *point* represents the mean hourly water loss (± 1 standard error) of ten animals. (C.S. Crawford and J.L. Cloudsley-Thompson)

the fact that his animals were young specimens maintained on moist tissue paper whereas ours were full-grown individuals captured in the desert and kept on moist sand in the laboratory. Evidently the wax layer of the vinegaroon, unlike that of desert scorpions, is easily abraded by dust and sand, and then becomes porous. In consequence, despite their comparatively large size and, for an arthropod, rather low surface to volume ratio, vinegaroons are compelled to hide away in moist micro-habitats from which they emerge only at night to hunt their prey.

The early work on DDT (dichloro-diphenyl-trichloroethane) was carried out with this substance in a powdered form, diluted with alumina dust (Al_2O_5). Much to the surprise of the investigators, the control insects, exposed only to the alumina dust, died before the experimental animals treated with DDT plus alumina dust. This was because, after the initial burst of activity stimulated by the insecticide, the experimental insects were less active than the controls, whose cuticles become abraded by the particles of alumina. Consequently they died as a result of dehydration before the experimental insects died from poisoning.

At this point, the reader may be wondering why the lipid barrier to the movement of water should be sited on the outside of the cuticle rather than near the epidermis where it would be better protected. A functional advantage of its actual position lies in the fact that the procuticle is, itself, more resistant to the passage of water when it is fully hydrated than it would be if dry. The same principle was employed by the manufacturerers of breakfast cereals in the days when waxed paper bags were used to keep their products dry. (Today they use plastic wrappers). Was the wax layer on the outside or inside of the paper? The answer is, on the inside. In this case, the cereals were packed in dry air and the function of the wax was to prevent the penetration of moisture. So it was more efficient for the paper to be in equilibrium with the damp atmosphere of the home, than with the dry air surrounding the cereals.

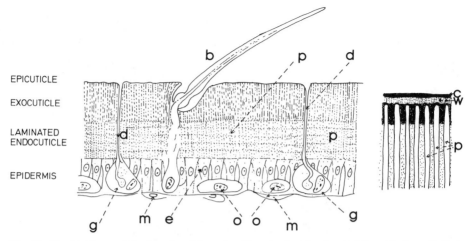

Fig. 10. The histology of the insect cuticle and epidermis, showing *b*, a bristle with the cells that secrete it; *p* pore canals; *g* dermal glands and their ducts; *e* epidermal cells; *o* oenocytes; *m* the basement membrane. *Right* detail of epicuticle (schematic) with cement layer. *c* wax layer; *w* wax layer with cuticulin and polyphenol layers underlying it and above the exocuticle with its pore canals. (After Wigglesworth 1964)

In addition to chitin, the arthropod cuticle is composed largely of 'arthropodin', a mixture of several proteins. Resilin is another protein constituent of the cuticle. It is rubber-like, with protein chains bound together in a uniform three-dimensional network. It provides the elasticity of the thorax of flying insects and is present in many other elastic structures.

2.5 Growth and Ecdysis

The possession of a hard cuticle renders moulting necessary if growth is to take place. Larval insects such as caterpillars may have loose-fitting and highly folded cuticles on their bodies which allow room for growth. Ecdysis of sclerotized structures, the head capsule and mandibles, for instance is, however, still necessary for growth because these cannot stretch after they have hardened.

Before moulting takes place, the epidermal cells separate from the cuticle and begin to lay down a new cuticle. At the same time, they secrete a 'moulting fluid' into the space between the new and old cuticles; this contains proteolytic enzymes and chitinase which digest the old cuticle. The new cuticle is not digested, however, because it is protected by a layer of cuticulin, which is the first layer of the new epicuticle to be deposited. As the cuticulin is laid down, the oenocytes, which had previously swollen greatly, now decrease in size. It is therefore believed that the secretions of these cells provide the raw material of the cuticulin layer.

As the procuticle develops, continuity between the epidermis and the epicuticle is maintained by the pore canals (Fig. 10). Through these pass the polyphenols and finally the epicuticular waxes. The latter are secreted as long filaments, and represent liquid/water liquid crystals. They exist in a number of phases, in the middle one of which the crystals have hydrophobe carbon chains directed towards the interior of the filament with hydrophil polar groups towards the liquid/water interface at the surface. In this way, insoluble lipid material passes through the water-impregnated cuticle. The cement layer is secreted by dermal glands, whose ducts discharge onto the surface of the wax layer. Finally, the insect emerges from its much depleted old cuticle, which splits along ecdysial lines of weakness, and swallows air so that its bulk is expanded. The presence of air in the crop, coupled with tonic contractions of the muscles of the body wall, creates considerable pressure in the haemolymph, which is distributed evely throughout the body, causing the soft and extensible new cuticle to expand. After this has taken place, the new cuticle begins to harden and darken in colour. The polyphenol tanning agents are apparently discharged from the tips of the pore canals and diffuse inwards so that sclerotization begins at the outer layer of the exocuticle and works inwards from there.

2.6 Respiration: Lung-Books and Tracheae

An arthropod completely covered with an almost impervious epicuticular wax layer would be in the fortunate position of losing very little water through transpiration, but it would not be able to breathe at all because the O_2 molecule is larger than that of H_2O. A respiratory mechanism has therefore been evolved which permits gaseous exchange without excessive water loss. The primitive respiratory organ of Metazoa is the skin, but special respiratory organs have been evolved in all but the smallest of the arthropods. In Onychophora, insects and myriapods, a system of spiracles, tracheae and tracheoles carries oxygen directly to the tissues where metabolic processes take place. In most arachnids, on the other hand, the chief respiratory organs are lung-books, directly modified from the gill-books of aquatic ancestors. These communicate with the external atmosphere by a small pore, and contain numerous respiratory lamellae (Fig. 11). Scorpions have lung-books only, while spiders are passing through a primitive lung-book stage from which none have yet emerged. Two pairs of lung-books, without tracheae, occur in the more primitive families, while most others have an anterior pair of lung-books and a posterior pair of tracheae. The respiratory significance of the latter is slight. Lung-books provide a localized respiratory area from which oxygen is distributed by the respiratory pigment haemocyanin in the blood. Insects possess no respiratory pigments in their haemolymph because all their tissues are supplied with oxygen directly by the tracheoles to cellular respiratory cytochrome pigments.

Tracheae have evolved secondarily in Solifugae where their presence may be related to an extremely active predatory life; but the same explanation has also been given for the presence of lung-books in *Scutigera* — while most other Chilopoda have tracheae and tracheoles! The rate of diffusion of oxygen through the tracheoles is probably one of the factors limiting the size of insects. Heavyweights, like *Goliathus regius* and *Megasoma elephas* (Sect. 2.1.1), are slow and sluggish. Other large insects tend to have long, slender bodies like dragonflies and mantids. The longest insect is probably the walking-stick *Pharnacia serratipes* (Phasmida) which may reach 33 cm in length. (The longest centipede is *Scolopendra gigantea* from tropical America which may grow

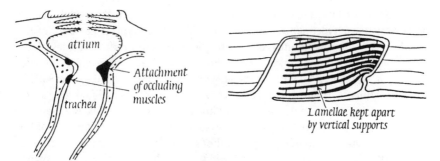

Fig. 11. Spiracle of a butterfly larva (*left*) and lung-book of a spider (*right*) as seen in section. Diagrammatic. (J.L. Cloudsley-Thompson after Wigglesworth 1964)

nearly to 30 cm). Tracheal respiration could, theoretically, be quite efficient, even in much larger arthropods, if there were sufficient air sacs to circulate tidal air through the main tracheal branches so that the distances to be covered by diffusion were not too great. Nevertheless it is one of the deterrents to larger size in arthropods.

The spiracles of insects and the lung-books of arachnids are normally kept closed by special muscles which relax only when the carbon dioxide content of the blood reaches or exceeds about 5%. In this way, the loss of water, which inevitably accompanies respiratory exchange, is kept to a minimum. (In a similar way, tortoises hold their breath when the temperature rises and take relatively few deep breaths. Under excessively hot conditions, however, water conservation becomes less important than body temperature, and the animals then pant).

Like the integument of the body, tracheae consist of cuticle (which is cast off at ecdysis) underlain by epidermal cells. They are strengthened by a spiral thread or 'taenidium' which continues into the finest branches of the tracheoles. In consequence of this, the tracheae and tracheoles are not easily compressed. Their gross structure resembles that of flexible metal gas tubing. They are usually circular in cross-section, which helps to prevent collapse but, in certain insects such as the water-beetle *Dytiscus*, they are elliptical and more easily compressed.

The conflict between the incompatible requirements of respiratory exchange and the prevention of water loss is illustrated by comparison of two common European spiders, *Amaurobius ferox* and *A. similis*, both of which have a cuticular wax layer with a critical temperature at about 30°C, above which they quickly lose water by evaporation in fry air. At lower temperatures, however, the rate of water loss in *A. ferox* is almost double that of *A. similis*. *A. similis* 'tires' very rapidly when forced to run at full speed without stopping, and is almost always overcome in fights between evenly matched individuals of the two species. *A. ferox* can run for longer before 'tiring' but, like all larger spiders, it can only run flat out until the oxygen stored in the haemocyanin of the blood is exhausted. After this, it begins to plod along rather slowly! Both species can, however, run for long periods when supplied with oxygen, and it appears that the greater stamina of *A. ferox* depends upon the presence of mor more laminae in the lung-books. The larger respiratory surface has been acquired at the expense of greater dependence upon environmental humidity. This is reflected in the distribution of the two species: *A. ferox* dominates in damper environments and on coastal islets, while *A. similis* is more plentiful in drier localities.

The question as to whether lung-books or tracheae and tracheoles are the more efficient cannot be answered, since comparative physiological studies are lacking. Nevertheless, the presence of localized respiratory organs necessarily entails a better developed circulatory system, with blood containing a respiratory pigment. Elaborate tracheae occur only in small spiders which have a large surface to volume ratio and are therefore more prone to desiccation. It would seem not unlikely that the most efficient system has been evolved in relation to the sizes and ways of life of different taxa of spiders.

2.7 Nutrition and Excretion

2.7.1 Nutrition

Woodlice and millipedes are almost entirely vegetarian, while centipedes and arachnids (with the exception of certain mites) are primarily carnivorous, but different species of insect seem to be able to thrive on almost any kind of organic matter that will support metabolism. Most of the larger arachnids, which do not inhabit damp leaf litter and similar cryptozoic habitats, are able to take up water from moist soil, provided that the humidity of the substrate is more than 12%. Furthermore, since they are predators, they obtain considerable amounts of liquid from the body fluids of their prey. Insects that feed on dry materials are faced with the greatest problem as far as water shortage is concerned.

The alimentary canal of insects is divided into three regions: the foregut (stomodaeum), the midgut (mesenteron), which is endodermal, and the hindgut (proctodaeum). The first and last of these are ectodermal in origin: consequently they have chitinous cuticular linings which are shed at each moult. In most insects, digestive enzymes are secreted by the salivary glands at the endodermal cells of the midgut. The oesophagus leads into a muscular pharynx which is frequently expanded to form a crop in which food is stored. In the Orthoptera, a considerable amount of digestion takes place in the crop, assisted partly by the enzymes from the salivary glands which are mixed with the food, and partly by digestive enzymes regurgitated from the midgut. In *Periplaneta*, it has been shown that absorption and assimilation of digested food may also take place in the crop.

Food passes from the crop, through the proventriculus or gizzard, into the midgut which, with its caecae, is the chief region of food absorption. In many insects, the delicate cells of the mid-intestine are protected from mechanical damage by a covering known as the 'peritrophic membrane' through which the digestive enzymes and their products pass freely. The peritrophic membrane probably has a function analagous to that of mucus in the vertebrate intestine. It is composed of chitoprotein, which resembles mucoprotein in many respects: it is absent from several species of insects with an entirely fluid diet. In most insects, the membrane is composed of loose concentric lamellae but, in Diptera, Isoptera and Dermaptera, a single uniform layer is secreted by a group of cells at the anterior end of the midgut. This is constantly renewed, portions of the older membrane often forming a wrapping round the faecal pellets as they are extruded.

In plant-sucking bugs (Homoptera) which feed on the copious watery sap of plants, a characteristic 'filter chamber' allows fluid to pass directly from the anterior to the posterior regions of the midgut, thus shortcircuiting the main digestive segment of the gut which receives only the more valuable constituents of the sap. Filtration is physiological, not merely physical. It permits the absorption of very large quantities of food, necessary because of the scarcity of nitrogenous material in plant sap, without unnecessary dilution of the digestive enzymes.

There is no phagocytosis in insects, where all the products of digestion are absorbed. Absorption and secretion are usually performed by the same midgut cells, but occasionally there is some division of labour. This is most marked in the tsetse fly *Glossina*, where the long coiled midgut is divided into three regions. In the anterior, water is absorbed and there are no enzymes; in the middle part enzymes are secreted and the food (blood) is blackened; in the posterior part absorption takes place. For an hour or two after feeding, tsetse flies excrete a clear liquid containing surplus water from their food. In Diptera, Lepidoptera and Hymenoptera, whose diet is usually liquid, the excreta is always liquid too, but even this is considerably concentrated as each bolus passes through the different regions of the midgut.

The beginning of the hindgut is marked by the junction of the malpighian tubules with the alimentary canal (Sect. 2.7.2). It receives excretory materials from these in addition to indigestible remains of the food, and stores them pending evacuation. The hindgut is the principal region of reabsorption of moisture from the excreta.

2.7.2 Excretion

The function of excretion and osmoregulation is the maintenance of a constant internal environment, and the physiology of nitrogenous excretion in terrestrial arthropods is dominated by the necessity for the economy of water. In consequence, their excretory compounds must be solid or highly concentrated. Like reptiles and birds, insects eliminate nitrogenous waste by excreting insoluble uric acid so that little or no water need be lost in the process. In a similar way, arachnids excrete guanine.

The evolution of uric acid metabilism is related to the development of a 'cleidoic' or enclosed egg surrounded by a relatively impermeable membrane or shell. Within such an egg, ammonia would soon accumulate and become toxic, while a concentration of urea would upset the osmotic relations of the developing embryo. Insects, like reptiles and birds, are therefore uricotelic, and develop the necessary machinery to excrete uric acid in the egg stage: they retain this throughout their lives. So, the answer to the question: 'which came first, the chicken or the egg?' should be 'the egg'! (Aquatic amphibians excrete ammonia as to the larvae of terrestrial forms, while the adults are ureotelic and excrete urea. Mammals have the advantage of an abundant supply of water during their embryonic development, and use urea as the end product of nitrogen metabolism throughout their life).

The excretory organs of Crustacea, maxillary and antennal or green glands, are modified coelomoducts secreting urea and ammonia: they have an important osmoregulatory function. Similar excretory coelomoducts occur in *Peripatus* and in Arachnida where they take the form of coxal glands. In the Oniscidae, over 50% of the soluble non-protein nitrogen is excreted in the form of ammonia. The level of nitrogenous excretion is appreciably lower in the more terrestrial woodlice, however, which suggests that adaptation to terrestrial life has been attended by a transformation of ammonia to other, less toxic, products (Chap. 3.4.2).

The chief excretory organs of Arachnida are the coxal glands already mentioned, but malpighian tubules also occur. The latter are not homologous with the malpighian tubules of insects. Several other types of cells and glands may also have an excretory function, and it has been suggested that the silk of spiders may have evolved from some substance originally excretory in nature. Malpighian tubules are present in both Chilopoda and Diplopoda although, again, homologues are doubtful. The sloughing of the gut wall or peritrophic membrane may also have an important excretory function in millipedes.

In Collembola and Thysanura, malpighian tubules are absent and the chief excretry organs are cephalic tubular glands similar to, but not homologous with, the antennal glands of Crustacea. Granules of uric acid are stored away as harmless accumulations in the fat bodies of Collembola and the malpighian tubules of cockroaches never contain uric acid, which again accumulates in urate cells of the fat body.

The histophysiology of excretion is best explained by reference to Fig. 12. In the first stage an alkaline solution of sodium and potassium urate is secreted at the distal end of the malpighian tubule. At the proximal end of the malpighian tubule, water is absorbed, the reaction becomes slightly acid, and uric acid crystals are precipitated. These then pass into the hindgut and more water is absorbed in the rectal gland. Finally, a dry faecal pellet containing uric acid crystals wrapped in peretrophic membrane is expelled to the exterior. There are numerous variations upon the theme outlined above. For example, in Orthoptera, Neuroptera and many Coleoptera, uric acid crystals separate only in the hindgut. In muscid flies and mosquitoes, which have liquid faeces, uric acid is precipitated throughout the tubules. Furthermore, other excretory compounds such as urea, amino acids, especially leucine, lime and calcium oxalate are occasionally present, whilst ammonia is produced by *Calliphora* and *Lucilia* larvae, which have a liquid diet. Nevertheless, Fig. 12 gives an indication of th basic excretory mechanism of insects and shows how nitrogenous wastes are eliminated without concommitant loss of water.

2.8 Ecological Considerations of Size

Natural selection can be defined as the survival of the most fit, with the inheritance of the factors in which the fitness lies. Successful individuals are those which survive long enough to be able to reproduce and bequeath genetic material to their offspring and subsequent generations. Even if an animal were to become so large and powerful as to be invulnerable to predatory enemies, it would not necessarily be successful. If success is to be evaluated in term of passing on 'selfish' genes to the next generation,

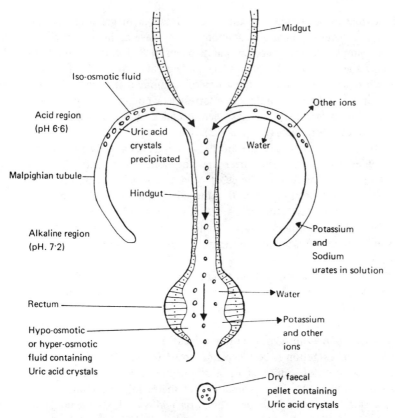

Fig. 12. Excretory processes in *Rhodnius*. (J.L. Cloudsley-Thompson after Wigglesworth 1964). Further explanation in the text

it might be more effective to produce a larger number of young which reach maturity more rapidly and with less expenditure of metabolic energy in so doing. Natural selection must at times favour economy, even at the expense of survival and security.

Not only are the size, shape, basic structure, body proportions, physiology, environment, food and mode of life of each kind of animal closely interrelated, but there is also a cybernetic, or feed-back, relationship with the sizes of its populations. It is commonplace that small animals are more abundant than large ones, codified in the concept of the pyramid of numbers. Large numbers of small herbivores at the bottom of the pyramid are preyed on by smaller numbers of larger predators at the top. This is why big, fierce animals are rare. At the same time, a large number of small animals can exploit a limited area more thoroughly than a small number of larger animals, but cannot travel so far in search of food. All these factors influence the relationship between size and form, in addition to the physical consequences of allometry and surface to volume ratio, in terrestrial arthropods.

Further Reading

Alexander R McN (1971) Size and shape. Edward Arnold, London (The Institute of Biology's Studies in Biology No 29)

Bereiter-Hahn I, Matolsty AG, Richards KS (eds) (1984) Biology of the integument, invertebrates. Springer, Berlin Heidelberg New York Tokyo

Bursell E (1970) An introduction to insect physiology. Academic Press, London New York

Cloudsley-Thompson JL (1977) The size of animals. Meadowfield, Shildon Co Durham (Patterns of Progress Zoology, Vol 3)

Dalingwater JE (1981) Chelicerate cuticle structure. In: Nentwig W (ed) Ecophysiology of spiders. Springer, Berlin Heidelberg New York Tokyo, pp 3–15

Ebeling W (1974) Permeability of insect cuticle. In: Rockstein M (ed) The physiology of insect, 2nd edn. Academic Press, New York San Francisco, pp 271–343

Edney EB (1977) Water balance in land arthropods. Springer, Berlin Heidelberg New York (Zoophysiology and Ecology, Vol 9)

Foelix RF (1982) Biology of spiders. Harvard University Press, Cambridge Mass London

Kerkut GA, Gilbert LI (1985) Comprehensive insect physiology biochemistry and pharmacology, vol 3. Integument respiration and circulation. Pergamon Press, Oxford New York Toronto Sydney Paris Frankfurt

Loveridge JP (1980) Cuticular water relations techniques. In: Miller TA (ed) Cuticle techniques in arthropods. Springer, Berlin Heidelberg New York (Springer Series in Experimental Entomology)

Neville AC (1975) Biology of the arthropod cuticle. Springer, Berlin Heidelberg New York (Zoophysiology and Ecology, Vol 4/5)

Richards OW, Davies RG (1977) Imms general text book of entomology, 10th edn. Chapman & Hall, London (2 Vols)

Wigglesworth VB (1964) The life of insects. Weidenfeld and Nicolson, London

Wigglesworth VB (1972) The principles of insect physiology, 7th edn. Chapman & Hall, London

3 The Conquest of the Land by Crustacea

3.1 Types of Adaptation

The ancestors of today's terrestrial arthropods probably followed more than one route when they first conquered the land. Some may have ventured across sandy beaches or the rocky intertidal zone, others went through mangrove swamps or by way of fresh water streams and lakes to the moist humus and leaf litter of tropical rain forests. In contrast, the vertebrates are believed to have emigrated to land via oxygen-deficient tropical swamps. The study of extant Crustacea illustrates some of the ways in which ancestral arthropods may have become adapted to life on land, and it is possible to gain some idea of the course of evolution by correlating the adaptations of existing species with the various environments they inhabit.

Crustaceans have become adapted to land through a number of morphological, physiological, biochemical, and behavioral modifications. Some of these are shared by all terrestrial forms, others are unique to the inhabitants of particular habitats. For instance, a species of land crab that lives for some of the time in wet mud has quite different physiological needs than one which burrows in dry grassland, or another that inhabits sandy beaches just above the high-tide level. Among the taxa of crustaceans having truly terrestrial members, the amphipods have achieved success on land primarily by behavioral means, whereas the isopods and decapods have exploited morphological, physiological and biochemical adaptations to a greater extent. Behavioural modifications ensure that water loss is minimal, and that environmental extremes are avoided; physiological adaptations are concerned with nitrogenous excretion, water balance, respiration and osmotic regulation. Furthermore, even closely related species of terrestrial crustaceans, such as land crabs, may have approached land by different routes.

3.2 Transition from Water to Land in Amphipoda

Completely terrestrial amphipods, as distinct from supra-littoral forms, are absolutely independent of standing water, for reproduction, feeding, distribution or general well-being. While there are degrees of adaptation to life on land, most species of terrestrial amphipods are almost as terrestrial as woodlice or myriapods. They are absent from Europe, much of Asia, North and South America (except for accidental introductions), but are found in Japan, the Philippines and the Indo-Malaysian region.

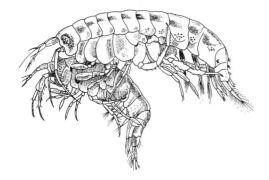

Fig. 13. Gammarids (*Gammarus*) mating. Male (*above*) and female (*below*). (After Della Valle in Smith and Weldon 1909)

Sandhoppers (Talitridae) are common on most shores throughout the world, except in the Arctic and Antarctic. They inhabit the region between the low tide level and the spray zone above high tide mark, sometimes occurring even further inland.

Terrestrial amphipods, or landhoppers, probably evolved from supra-littoral species, and were able to enter leaf litter in tropical and southern cold-temperate forests without ever passing through a fresh or brackish water stage.

In many ways, the greatest problems in the transition from sea to land must have been overcome when talitrids emerged from the subtidal regions and entered the supra-littoral zone. The substitution of air for sea water requires both mechanical and physiological adaptation. The method of locomotion changed from swimming to jumping: this was achieved by shortening and strengthening the limbs and uropods. The animals balance on their third pereopods while turning the abdomen under the body so that the ends of the uropods 3 and the telson press into the sand. When the abdomen is flexed and straightened out, sandhoppers and landhoppers are propelled into the air. In landing, uropods 1 and 2, which are strong and spiny, are used as shock-absorbers and act as levers. Lateral body shields have been developed from the side and epimeral plates of the thoracic and abdominal segments, and these protect the gills, eggs and developing young, perhaps also preserving moisture in the brood pouch. Many marine genera also possess well-developed lateral shields so sandhoppers were, in this respect, pre-adapted for life on the sea-shore (Fig. 13). Modification was also necessary for respiration, excretion and reproduction on land. There is a trend away from thick heavily-spined appendages to slender, finely-spined legs, antennae and mouthparts. Reduction and loss of the pleopods is also a striking feature of terrestrial amphipods.

Forest leaf mould supports a prolific cryptozoic fauna, which includes amphipods and isopods, as it is humid, insulated, and contains abundant food. The high relative humidity restricts evaporation to a minimum, and water loss can be replaced by dew and rain. The number of eggs carried by terrestrial and supra-littoral amphipods ranges from 1 to about 24, compared with 50–225 eggs carried by marine species. The eggs are retained within the brood pouch, where development, in terrestrial forms, is favoured by the presence of free water. These adaptations to life on land appear to have continued trends that were already existing in aquatic amphipods.

Like terrestrial isopods, landhoppers migrate to deeper layers during periods of drought, and may climb trees and bushes to avoid drowning during heavy rain and floods. Problems resulting from a change in diet include those relating to the uptake of heavy metals which, in plant tissues, are firmly bound to organic complexes. Terrestrial amphipods, like woodlice, have a greater capacity for the storage of copper than have marine species; the copper is more rigorously compartmentalized, and its movement around the body more strictly regulated. Decapods, too, are able to tolerate large quantities of copper, which is fixed in the exoskeleton and discarded when moulting.

The Amphipoda have not achieved the degree of terrestrial independence found in the Isopoda, and are restricted to a fairly narrow niche in forest leaf mould, or in grassland where conditions are somewhat similar. Although more emancipated than the terrestrial Decapoda in that they are able to reproduce on land, terrestrial amphipods are even more restricted to moist habitats.

3.3 Transition from Water to Land in Decapoda

Terrestrial and semi-terrestrial representatives of the order Decapoda are found mainly in tropical and subtropical regions of the world. They are included among four families of Macrura, one of Anomura and seven of Brachyura. With the invasion of the intertidal zone and, subsequently, of the land above the high tide level, some species evidently found it advantageous to develop new patterns of courtship and reproduction, while returning to the sea or some other body of water in which their young might hatch. These species also found it necessary at times to withstand, or evade, wide variations in temperature and relative humidity while simultaneously maintaining a normal balance of salts and water. Many of the separate problems accompanying the transition from water to land may thus be considered as facets of the two major ones: continuance of reproduction and early development, and regulation of temperature, salts and water.

Female aquatic decapod crustaceans are known to produce pheromones to which mature males of the same species are sensitive. For instance, females of the Indo-Pacific marine crab *Portunus sanguinolentus* release a chemical that induces display and searching behaviour in males. These walk about on the tips of the legs with their bodies well elevated and claws extended. On touching a female that is about to shed her shell, the male seizes her and pulls her into a pre-copulatory position.

Among semi-terrestrial and terrestrial crabs, a soluble chemical sex attractant may have limited value, for even semi-terrestrial species mate on land after the tide has ebbed. Some species, particularly of hermit crabs, appear to have a courtship that involves the release of an air-borne pheromone to which the chemosensory hairs of the other sex may respond. Most semi-terrestrial and terrestrial crabs, including fiddler crabs, however, indulge only in visual displays during their courtship.

Terrestrial decapods usually mate while their shells are hard. Marine species produce large numbers of small eggs during spawning, as do semi-terrestrial and terrestrial species that return to the sea to spawn, but many fresh-water species produce small numbers of large and relatively impermeable eggs containing generous quantities of yolk. Just as the number and size of the eggs produced in a single spawning are related to the salinity of the water in which spawning occurs, so also is the type of larval development. Among species that reproduce in fresh water, development usually takes place within the egg, and the young hatch as miniatures of their parents.

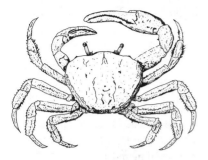

Fig. 14. Land-crab (*Gecarcinus*) showing lack of conspicuous morphological adaptations to terrestrial life

The water and heat relationships of fiddler crabs (*Uca* spp.) at Inhaca, Mozambique, are related to the extent of their terrestriality. For instance, the species *U. urvillei, U. marionis, U. chlorophthalmus, U. annulipes* and *U. inversa,* generally speaking, respectively occupy progressively more terrestrial habitats. Their upper lethal temperatures are sharply defined and range from 40.0°C for *U. urvillei* to 43.3°C for *U. inversa.* [*U. marionis* is apparently aberrant in having an upper lethal temperature about 41.4°C which falls between those of *U. annulipes* (41.2°C) and *U. chlorophthalmus* (40.8°C)]. Upper lethal temperatures are about 2°C lower for each species in September than in January. The rates of water loss through transpiration stand in the same order, being greatest in *U. marionis* and *U. urvillei* and least in *U. annulipes.* (*U. inversa* is anomalous as it loses water fastest of all when first exposed to dry air although, after half an hour or so, its rate of loss becomes least of all). Species which transpire faster tend to have lower body temperatures than those more terrestrial species which transpire more slowly. These measurements are consistent with observations recorded in the natural environments of the different species.

When decapod crustaceans abandon the sea in favour of the intertidal zone, they move into an environment in which salts and water are relatively scarce. Terrestrial crabs, such as *Gecarcinus lateralis* (Fig. 14) only have air within their branchial chambers, while semi-terrestrial genera, such as *Uca* and *Ocypode,* regularly carry small amounts of water around their gills. They also have fewer gill filaments with less surface area than the gills of aquatic species. Some terrestrial crabs, e.g. coconut crabs (*Birgus:* Anomura), have reduced gills and supplementary lungs in the branchial chamber. They are able to take up water from a moist substratum and store it in the pericardinal sacs. This is particularly useful at ecdysis when water is needed to stretch the new soft cuticle before it hardens.

Crabs achieve a proper balance of salts and water by drinking selected amounts of sea and rain water or, as in the case of terrestrial hermit crabs, by filling their shells with water of the desired salinity. In general, terrestrial and semi-terrestrial species are strong hyp-osmoregulators and, to a lesser extent, hyper-osmoregulators as well. Their antennal glands function in ionic regulation and the gills and alimentary canal also take part in this process. In the more terrestrial species of decapod crustaceans, emancipation from water is almost complete, save for the need, when spawning, to return either to the sea or to fresh water. Since terrestrial amphipods and isopods can live and breed entirely on land, decapods are less emancipated in this respect. In physiological terms, however, particularly with regard to their resistance to desiccation and their capacity to mobilize ions and water, the decapods are well advanced in their invasion of the land.

3.4 Transition from Water to Land in Isopoda

Terrestrial isopods constitute the suborder Oniscidea, of which the earliest fossils occur in Baltic amber from the upper Eocene or lower Oligocene. These fossil forms are congeneric with the extant genera *Trichoniscus, Porcellio* and *Oniscus,* and therefore give no idea as to the path by which their ancestors conquered the land. On the other hand, by correlating the morphology, physiology, and behaviour of existing species with their various environments, it is possible to gain some insight into the ways in which this may have taken place.

3.4.1 Morphology

On grounds of comparative morphology, there is considerable evidence that the Oniscidea are polyphyletic. One group, represented by the genera *Tylos* and *Helleria,* resembles marine valvifers such as *Idotea.* A second group includes *Trichoniscus* and other Trichoniscidae (and the primitive cave-dwelling genus *Cantabroniscus*). A third group contains all the remaining Oniscidea. The second and third groups, according to A. Vandel, are derived from undetermined but distinct sources among the marine isopods. A revised classification has been outlined by Sutton and Holdich (1984).

The ancestors of woodlice probably became terrestrial during the second half of the Palaeozoic era. This conclusion is based on the fact that all the main types of organization within the Oniscidea show a world-wide distribution and, consequently, must have a very ancient origin. Although the Oniscidea include the most specialized of the Isopoda (apart from parasitic forms), they represent an evolutionary cul-de-sac. The most primitive of the woodlice and, at the same time, the least well adapted to terrestrial conditions are littoral species belonging to the family Ligiidae. The family Trichoniscidae also occurs in very moist places; but the Oniscidae, Porcellionidae and Armadillidae are able to survive for progressively longer periods of time in drier environments (Fig. 15). This sequence is also one of increasing morphological specialization, which can be seen in the following major trends:

a) reduction in size (from 1–2 cm to a few mm in length);
b) increase in the number of spines and tubercles;
c) evolution of the ability to roll into a ball;
d) appearance of respiratory pseudotracheae.

There are also progressive changes in the morphology of the head, body and appendages, whose adaptive significance is less clearly apparent, and of the internal anatomy.

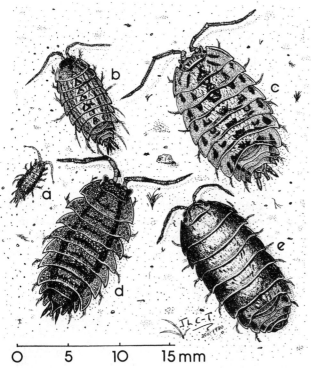

Fig. 15a–e. European woodlice: a *Trichoniscus pusillus;* b *Philoscia muscorum;* c *Oniscus asellus;* d *Porcellio scaber;* e *Armadillidium vulgare.* (J.L. Cloudsley-Thompson)

3.4.2 Physiology

The most important physiological adaptations of Isopoda to life on land are concerned with the conservation of water. The main sources of water loss in Arthropoda are through cutaneous transpiration, moulting and excretion. Not only is the arthropod cuticle an exceedingly complex structure, which varies between one taxon to another, but information about cuticular water loss may be expressed in various ways. These are not strictly comparable, and each of them is appropriate to a particular type of enquiry. E.B. Edney (1968) has summarized them as follows:

a) as a proportion of the total weight of the individual concerned. This parameter can be useful when studying limits of tolerance, but tells little about relative permeabilities of integuments;

b) as a rate of transpiration per unit of surface area. This is satisfactory for comparative purposes, but tells little about the probable biological effects unless the size of the animals is known;

c) as a rate of transpiration per unit area and vapour pressure difference. Again, this permits calculation of weight-specific loss only if the surface/weight coefficient is known.

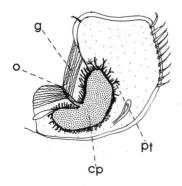

Fig. 16. Exopodite of the first pleopod of *Porcellio* showing pseudotracheae. (After K.W. Verhoeff in Little 1983), *cp* cavity of pseudotracheae; *g* grooves on anterior border of pleopod; *o* opening of pseudotrachea; *pt* pseudotracheal tubules

Although some research workers claim to have obtained evidence to the contrary, in general transpiration through the integument of land isopods, however it is expressed, tends to be proportional to the vapour pressure of the surrounding air. Within the biological range, temperature does not greatly affect the permeability of the cuticle. This does not appear to possess a water-proofing mechanism consisting of an oriented layer of polar lipid molecules in the epicuticle, as is characteristic of insects and arachnids (Chap. 2.4) (Fig. 8). Loss of cuticular water varies greatly between species, but there is a general relationship between rate of transpiration and habitat. Thus the rate of water loss in mg/cm^2/h/mm Hg $\times$ 10^3 at 30°C is 220 from the sea-slater *Ligia oceanica*, 180 from *Philoscia muscorum*, 132 from *Oniscus asellus*, 110 from *Porcellio scaber* and 85 from *Armadillidium vulgare*. The figure for the North African and Middle Eastern *Hemilepistus reaumuri* (Porcellionidae), under comparable conditions, is only about 23, while that for the North American desert species *Venezillo arizonicus* (Armadillidae) is as low as 15. From consideration of these figures, it is obvious that one aspect of the adaptation of woodlice to life on land involves reduction in cuticular transpiration.

Oxygen uptake by land isopods occurs mostly through the pleopods. These are simple gill-like structures in the Ligiidae, whereas in families better adapted to terrestrial life they are modified by invaginations into pseudotracheae (Fig. 16), as mentioned above. Water loss is reduced in forms possessing these respiratory modifications, but the way in which this is achieved is not yet known. The advantage to insects of a tracheal system lies in the presence of occlusible spiracles which permit water loss to be reduced to a minimum, but similar spiracles are not found in woodlice. Nevertheless, the respiration of land isopods is remarkably efficient.

The water-conducting system is a very characteristic feature of all Oniscidea. It carries fluid to the 'pleoventralium', the space between the ventral body wall and the exopodites of the pleopods in which are situated the plate-like endopodites. Opinions differ both as to the origin of this fluid and to its fate. It probably originates from the maxillary excretory nephridia and is reabsorbed by being taken in at the anus, a process adding considerably to the rate of water loss by evaporation. The epithelium of the endopodites shows considerable modification, suggesting a transporting function: there are deep infoldings, which are not usually found under very thin cuticle. The basal plasma membrane is also highly convoluted, and its cells contain numerous mitochondria. It is well established that terrestrial isopods are

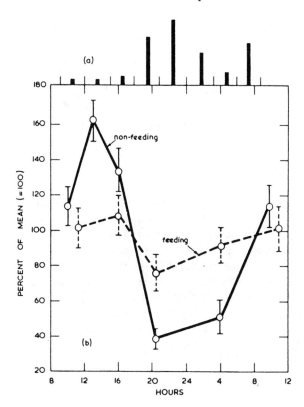

Fig. 17. a Activity levels of *Porcellio scaber* at different times of days; b levels of ammonia production at different times of day in feeding and non-feeding specimens of *P. scaber*. (Cloudsley-Thompson 1977, after W. Wieser, G. Schweitzer, R. Harterstein)

ammoniotelic, discharging nitrogenous waste products in the form of gaseous ammonia. This is excreted in solution by the maxillary nephridia and passes, in bursts, into the ventral conducting system. As the urine flows along the ducts of the system, it becomes more alkaline, which leads to the progressive volatilization of ammonia. Thus, the water-conducting system acts as an extension of the kidney, and the excretion of nitrogenous end products is closely linked with water economy — as is the rule in ureotelic and uricotelic forms, but not usually in the case of ammoniotelic animals. Consequently, woodlice are able to regulate the water contents of their bodies, to synchronize water flow, excretion rates, etc., and to discharge nitrogenous end products of metabolism with minimal expenditure of energy and water (Fig. 17).

The Isopoda have further adapted protein metabolism to terrestrial conditions by excreting mainly during periods of inactivity, when the animals are in their moist retreats and are in the least danger of losing too much water along with the ammonium carbonate that passes through the epithelium of the pleopod endopodites. As carbon dioxide is liberated into the atmosphere, the ammonium carbonate dissociates, releasing gaseous ammonia.

Loss in weight through transpiration does not immediately result in a corresponding increase in the concentration of the haemolymph. Indeed, in *P. scaber,* the haemolymph may remain constant during several hours of desiccation. This is due to the fact that water is taken up from the muscles of the body while salts are withdrawn into the alimentary canal. Later, as dehydration proceeds, wide osmotic variations in the haemolymph can be tolerated.

The upper lethal temperatures of woodlice in air saturated with water, over periods of 24 h, have been assessed as follows: *L. oceanica,* 29°C; *Ph. muscorum,* 30.5°C; *O. asellus,* 31.5°C; *P. scaber,* 36°C; and *A. vulgare,* 37.5°C. Lethal temperatures, therefore, vary according to habitat, but are affected by body size, the relative humidity of the air, the permeability of the cuticle, and by previous thermal history. Evaporative cooling is sometimes of survival value for short periods but is not effective for very long because the rate of dehydration is so high. Metabolic rate is affected by season: long photoperiod, as well as comparatively high temperatures, are necessary for reproduction.

Other physiological adaptations of terrestrial isopods include the ability to exploit a wide range of nutrients by varying the rate of ingestion and absorption of different kinds of food. In this way, an optimal composition of nutrients and energy is achieved from mixtures of foods, while coprophagy improves the nutritional value of primary food substances. Large amounts of copper can be stored in special 'cuprosomes' or granules of copper, sulphur and calcium within the cells of the hepatopancreas, and woodlice are able to select between types of food with different amounts of copper in them. In consequence, they have a low sensitivity to high concentrations of toxic substances in their food. The young are retained in the brood pouch during their early development when they are most sensitive to adverse conditions.

3.4.3 Behaviour

Poorly-equipped as most woodlice are, compared with insects and arachnids, to withstand desiccation on land, their behavioral mechanisms ensure that extremes of drought and heat are avoided by aggregating, for much of the time, in damp, dark places from which they emerge at nightfall when the temperature drops and the relative humidity of the atmosphere increases. The orientation reactions by which this is achieved are mainly *orthokineses,* which are simple effects on the rate of locomotion depending on the intensity of the stimulus, and *klinokineses,* in which the frequency of turning, or the rate of change in the direction of movement is again dependent upon the intensity of stimulation. Responses to light are *tropotaxes,* movements whose direction is dependent on a simultaneous comparison of the intensities of illumination stimulating the two eyes.

Woodlice show circadian rhythms of locomotory activity, and a fall in the intensity of the response to humidity after dark enables the animals for a time to walk in places drier than their daytime retreats. At the same time, increased photo-negative behaviour after exposure to darkness ensures that they return to cover promptly at daybreak and this, no doubt, helps them to avoid predatory birds. On the other hand, a reversal from negative to positive phototaxis is correlated with water loss by evaporation so that, should their day time shelters dry up, the animals are not trapped there by the light, but are able to wander in the open until, by chance, they find some other damp environment. Here they become rehydrated and their negative responses to light reassert themselves. The degree of nocturnal activity in different species is correlated with the ability to withstand water loss by transpiration. Woodlice do not, however, come out on windy nights when evaporation would be much increased. Seasonal changes also occur in the intensity of their humidity reactions, which shows a rise in spring, when the rains bring them out of hibernation. At the same time, seasonal changes also occur in distribution, and some species move from winter habitats under stones and litter to dead wood, or even climb up the trunks of trees.

Thanks to its behavioral adaptations, the desert woodlouse *Hemilepistus reaumuri* is one of the most successful herbivores and detritivores of the macrofauna in many arid areas of North Africa and the Middle East. It survives by digging deep burrows in spring, when the soil is damp, without which it would be unable to avoid desiccation in summer. These burrows have to be defended continuously against competitors, however, and the adults form monogamous co-operating pairs and later, with their progeny, strictly closed family communities. Entrance to the burrows is permitted only to individuals marked with the appropriate genetically determined pheromone. On this is based the communication system of the family group. (Some species of *Porcellio* have also evolved a social system, but they are not monogamous and their burrows are not defended).

When foraging, *Hemilepistus* spp. often make extensive excursions, returning to their burrows by the shortest possible route. They do not use landmarks to find their way home but, instead, keep track of the position of the burrow relative to their own position. During the homeward journey the woodlice navigate by the sun and, if this is obscured, use the pattern of polarization of the light in the sky. On return to the approximate position of their burrow, they search for its systematically, finally identifying the entrance with their antennal chemoreceptors. It is only by means of such complex patterns of behaviour that *Hemilepistus* spp. are able to exploit such extreme terrestrial environments.

Of the sense organs of woodlice, tricorns, and the antennal and uropodal spikes, are unique: they are probably concerned with terrestrial adaptations of behaviour.

3.5 Conclusion

As a group, crustaceans are generally ill-equipped for life on land, and most terrestrial forms survive either by remaining close to aquatic habitats, or by occupying moist micro-habitats. They have not become specialized for any restricted food substances: in consequence the ecological niches they exploit are relatively broad. Whereas the majority of terrestrial animals appear to have reached land via fresh water, crustaceans have made a more abrupt transition by way of the marine supralittoral zone. This is indicated by the fact that terrestrial genera of amphipods and isopods include supralittoral species, while fresh-water genera are more closely related to marine forms than to supralittoral and terrestrial species. In contrast, as we have already seen, the ancestors of the larger terrestrial chelicerates probably reached the land through tropical swamps, while smaller forms and the ancestral symphylan stock – from which the insects subsequently evolved – went by way of the moist soils of tropical rain forest. This hypothesis accords well with what is known of the behaviour and environmental responses of myriapods, Collembola, Thysanura and other Apterygota.

Further Reading

Bliss DE (1968) Transition from water to land in decapod crustaceans. Am Zool 8:355–392

Bliss DE, Mantel LH (1968) Adaptations of crustaceans to land: a summary and analysis of new findings. Am Zool 8:673–685

Cloudsley-Thompson JL (1977) The water and temperature relations of woodlice. Meadowfield Press, Shildon Co Durham (Patterns of Progress Zoology, Vol 8)

Edney EB (1954) Woodlice and the land habitat. Biol Rev 29:185–219

Edney EB (1960) Terrestrial adaptations. In: Waterman TH (ed) The physiology of Crustacea, vol 1. Acedemic Press, New York, pp 367–393

Edney EB (1961) The water and heat relationships of fiddler crabs (*Uca* spp.). Trans R Soc S Afr 26:71–91

Edney EB (1968) Transition from water to land in isopod crustaceans. Am Zool 8:309–326

Edney EB (1977) Water balance in land arthropods. Springer, Berlin Heidelberg New York (Zoophysiology and ecology, vol 9)

Hurley DE (1968) Transition from water to land in amphipod crustaceans. Am Zool 8:327–353

Little C (1983) The colonisation of land. Origins and adaptations of terrestrial animals. Cambridge Univ Press, Cambridge London

Powers LW, Bliss DE (1983) Terrestrial adaptations. In: Vernberg FJ, Vernberg WB (eds) Environmental adaptations. The biology of Crustacea, vol 8. Academic Press, New York London, pp 271–333

Smith G, Weldon WFT (1909) In: Harmer SF, Shipley AE (eds) Cambridge Natural History, vol 14. Macmillan, London

Sutton SL, Holdich D (eds) (1984) The biology of terrestrial isopods. Clarendon Press, Oxford (Symposia of the Zoological Society of London No. 53)

Størmer L (1979) Arthropods from the lower Devonian (lower Emsian) of Alken an der Mosel, Germany. Part 5. Myriapoda and additional forms, with general remarks on fauna and problems regarding invasion of land by arthropods Senckenberg lethaea 57:87–183

Vandel A (1966) Sur l'éxistence d'Oniscoides très primitifs menant une vie aquatique et sur le polyphylétisme des Isopodes terrestres. Ann Speleol 20 (1965):489–511

4 Insect Phylogeny and the Origin of Flight

Time spent on insect phylogeny is often wasted. The origins of most insectan orders is a mystery, and is likely to remain so into the foreseeable future — certainly until existing fossils have been studied more extensively and further palaeontological evidence has been collected. Nevertheless, some intriguing evidence regarding the possible interrelationships of the various orders of insects is available, and this will be reviewed in the present chapter.

4.1 Ancestry of Insects

Five major theories regarding the origin of insects have been proposed:

a) F. Brauer (in 1869) suggested that *Campodea* is close to the ancestral stock of the insects, and can, itself, be derived from chilopod ancestors. The first of these hypotheses is no longer generally accepted, while the second has been rejected on a number of occasions.

b) A.S. Packard (in 1873) postulated that insects might have a common origin with the order of small myriapods which, 7 years later, was named Symphyla by J.A. Ryder to mark 'the singular combination of myriapodous, insectan and thysanurous characters which it presents'.

c) H.J. Hansen (in 1883) claimed that the insects evolved from Crustacea. This view was subsequently elaborated into a wider theory of arthropod phylogeny by E.R. Lankester (in 1904) and, in various forms, afterwards received support from G. Carpenter, R.E. Snodgrass, and others.

d) A. Handlirsch (in 1908), from a study of fossil forms, suggested that insects might have evolved from trilobites, with a corollary (also in Lankester's theory) that the original insects were winged.

e) R.J. Tillyard (in 1930) demonstrated that both crustacean and trilobite hypotheses were untenable. He almost accepted the symphylan theory, rejecting it, however, only on what was then believed to be a fundamental difference between the progoneate Symphyla (whose genital ducts open in the anterior region of the body), and the opisthogoneate insects (whose gonads open posteriorly). He countered this problem by proposing an hypothetical 'protapteron' (analogous with the nauplius larva of Crustacea and like it in having few segments) as a common ancestor for the myriapods and insects.

Tillyard's views were criticized strongly for a number of reasons, not least because they depended upon E. Haeckel's theory of recapitulation, and consequently fell into disrepute. Ten years later, however, O.W. Tiegs solved the problem, using embryological evidence. He studied the development of an Australian symphylan, *Hanseniella agilis,* and showed that the gonoduct was a secondary epithelial invagination. From this, he deduced that the distinction between progoneates and opisthogoneates is not a fundamental one. Consequently, insects could well have been derived directly from symphylan ancestors without the necessity to postulate an ancestral protapteron to link the two taxa. Tiegs added further evidence indicating links between the Onychophora and Symphyla, and between the Symphyla and Protura. In 1947, even further evidence was obtained from a study of the Pauropoda. This work was summarized by Tiegs and Manton (1958).

Taking everything into account, there seems to be little room for doubt but that the insects evolved from terrestrial ancestors which had antennae; endognathous, labiate mouthparts; tracheal respiration; and, probably, 14 post-cephalic segments; 12 pairs of legs; coxal styles; eversible vesicles on most segments; and a pair of terminal cerci. Despite this, there are two difficulties in accepting the Symphyla as being really close to these hypothetical ancestors: first, many of the myriapodan characters they exhibit are derived rather than primitive and, second, they lead more directly to the Diplura than to the Apterygota. These secondary difficulties were largely resolved by Manton (1964) from a detailed comparative study of arthropodan mandibular mechanisms. Her conclusions were as follows: (i) The terrestrial onychophoran-myriapod-hexapod evolutionary line of organisms with jaws (i.e. mandibles) derived from whole limbs was completely separate from the crustacean and trilobite-chelicerate lines with jaws derived from the bases of the limbs. (ii) There was an early division of the animals with whole-limb jaws into two lines: one with unsegmented jaws and a rolling biting mechanism, from which the hexapods evolved; and the other with segmented jaws and a transversely biting mechanism, from which the myriapods evolved. (iii) The jaw mechanisms of Symphyla are of the myriapod type, and they therefore could not be relics of stock from which the hexapods had also evolved. Their resemblance to some primitive hexapods is due, partly to retention of primitive (*plesiomorphic*) characters that were possessed by their common ancestors, and partly to parallel evolution of convergence, as with the fusion of the 2nd maxillae into a labium. (iv) The unsegmented rolling jaws of *Petrobius* (Archaeognatha) are not far removed from the primitive type, from which the entognathy of the Collembola, Protura and Diplura could have evolved on the one hand, and ectognathous, secondarily transversely biting jaws of the Thysanura and winged insects on the other. The presumed evolutionary relationships between mariapods and hexapods are shown in Fig. 18. Needless to say, Manton's conclusions have by no means been universally accepted; indeed, many of them have been criticized severely.

4.2 The Origin of Wings

In contrast to the apterygote orders, pterygote insects are not basically cryptozoic. They possess an epicuticular wax layer (Chap. 2.2) and are able to tolerate desiccating conditions — a necessary pre-requisite to an aerial existence. So little is known about

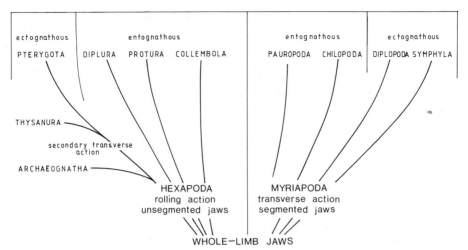

Fig. 18. The presumed evolutionary relationships of the hexapods and myriapods. (Based on S.M. Manton)

the origin of wings and flight in insects that any discussion of the problem must, of necessity, be almost entirely speculative. Indeed, few events in the evolution of animals have engendered as many contradictory hypotheses as has the origin of insect flight.

Since the venation of all insect wings can be homologized, it may be assumed that the Pterygota must have arisen from a common ancestor – which might well have been in existence in the early Devonian period or perhaps even in the Silurian – at least 100 millions years before the Paleodictyoptera of the Carboniferous and Permian periods. The earliest winged fossils are Middle Carboniferous (ca. 320 million years BP). Wings evidently originated early in the evolutionary history of the insects – perhaps towards the end of the Devonian (ca. 345 m. y. BP) – and, since they have been retained almost universally by the class, must, from the onset, have served an important function. Acccording to V.B. Wigglesworth (1976), that function was dispersal. Dispersal, and notably aerial dispersal, is characteristic of the inhabitants of temporary habitats. There may well have been times in the early Devonian or Silurian periods when desert conditions, with temporary habitats, were widely prevalent. In such environments there would be a high premium on the ability to undertake aerial migrations. The major alternative explanation is that the original function of wings was to facilitate escape from enemies. According to this, falling or jumping arboreal insects used 'pro-wings' first for altitude control in parachuting, then progressively for gliding or powered flight.

4.2.1 Aptelota and the Ancestry of Spiders

Before discussing the concomitants of these hypotheses, I would like to mention briefly a more fanciful idea, based on the alternative theory, which was proposed by

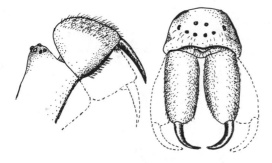

Fig. 19. Jaws in action in (*left*) a mygalomorph spider in which the prosoma is raised with fangs extended and the prey then impaled with a downward motion and (*right*) an araneomorph spider in which the chelicerae pivot sideways towards one another. (After R. and K. Preston-Mafham 1984. Spiders of the world. Blandford Press, Poole Dorset)

W.S. Bristowe (1958). Very early in their history, probably during the Carboniferous period, torsian of the basal segment of the chelicerae took place in some spiders, so that the fangs moved laterally inwards, and towards one another. These spiders were the ancestors of the suborder Araneomorphae. In the more primitive suborders, Liphistiomorphae and Mygalomorphae, the chelicerae strike forwards and downwards (Fig. 19). According to Bristowe, this change in the chelicerae may have been correlated with a departure from the ground-living habit, after insects had begun to hop and fly. When insects took to the wing, spiders evolved silk snares which trapped them in flight and on which the pick-axe blows of mygalomorph spiders would have been less effective than on solid ground or against the branches of tree-ferns. 'Is it unreasonable to suggest', asked Bristowe, 'that the spider menace represented a principal factor in the evolution of the hopping and flying powers which innumerable insects now possess? We all remember the riddle: "Why did the fly fly?" The answer: "Because the spider spied her" takes on a new significance if we accept the idea that flies might never have acquired the ability to fly at all had the spiders not existed!' Intriguing though it maybe, we cannot readily accept this ingenious idea — simply because insect wings appear to have evolved so long before the appearance of araneomorph spiders. At the same time, fossil evidence is so slight that it may nevertheless concern a germ of truth.

4.3 Paranotal Theory

It is widely, but not universally believed, that insect wings arose from mesothoracic and metathoracic paranotal lobes, homologous with those which occur in the orders of the Palaeozoic Palaeodictyoptera and of some Protodonata, Protorthoptera and Permian Ephemeroptera. Similar lobes are to be seen in the prothoracic shields of certain other Protorthoptera and of Blattodea, and in the abdomens of both adult and nymphal Palaeodictyoptera, Protorthoptera and Blattodea. When paranotal lobes occur in existing insects such as the Peloridiidae (Hemiptera), larval Ephemeroptera and Coleoptera, they are probably secondary: they serve functions such as the elimination of shadow in camouflage, protection from predatory attack, or streamlining. In ancestral forms, however, it is supposed that paranotal lobes served first as parachutes and then as aerofoils for gliding. Those in the mesothorax and metathorax became enlarged and modified. Changes took place in the skeleton, musculature and nerve supply of the thorax until, eventually, true flapping flight evolved. The prothoracic

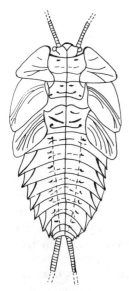

Fig. 20. Young nymph of the Palaeozoic terrestrial palaeodictyopteran *Rochdalia parkeri* illustrating the 'articulated' thoracic lobes. (After Wootton 1976)

paranotal lobes became obsolete and finally disappeared. Prothoracic paranotal lobes were most wing-like in the Palaeodictyoptera. These insects played an important part in A. Handlirsch's theory of the origin of insects from trilobites (Sect. 4.1). Although this theory is now completely rejected, the concept of the Palaeodictyoptera as being ancestral to all Pterygota was widely accepted for many years.

Although most entomologists probably adhere to the paranotal lobe theory — which postulates that insect wings have been derived from non-articulated lateral extensions of the thoracic tergites — study of the juvenile wings of Palaeozoic insects suggests that the 'pro-wings' of their nymphs were articulated structures which secondarily lost mobility by fusion with the tergum. They thus became evolutionarily convergent with paranotal lobes (Fig. 20). A second controversy has centred around their possible functions. Some authorities have suggested that pro-wings may originally have served to cover the spiracular openings or gills of amphibious ancestors, or that, like paranotal lobes, they protected and concealed their owners from predators. Others have speculated that they may have served to facilitate passive aerial dispersal by small insects or have aided in sexual displays. Yet another possibility is that they may have been an adaptation for behavioral thermoregulation. Because activity is dependent upon body temperature, any thoracic structure that enabled the leg muscles of ancestral pterygotes to heat rapidly would have enhanced locomotory efficiency and, consequently, survival.

The selective advantage of the paranotal lobes for thermoregulation does not exclude other possible functions which might well have been evolved concurrently. The 'spiracular flap' theory helps to explain why wing buds are always found near and above the spiracles. The 'gill-cover' theory probably provides the most acceptable reason for the rapid evolution of small pro-wings, once some of the ancestral insects had become aquatic (see below), both through the generation of respiratory currents and for their position above the spiracles. An aquatic origin for wings is also implied in the 'fin theory', according to which the little pro-wings, which would have had little or no effect in air, might have been highly useful as fins for propelling their owners through the water. Wigglesworth also introduced a novel hypothesis suggesting that wings originated from the coxal styles of Apterygota, which he assumed to be homologous with the abdominal gill-plates of mayfly nymphs. Although this homology has been disputed, there is, nevertheless, a not inconsiderable amount of other evidence supporting the 'stylus theory'.

In opposition to the hypothesis of an aquatic origin of wings, the following facts have been adduced. Almost all living, generalized insects have normal, functional spiracles in their immature stages, so these must also have been present in the juvenile stages of their ancestors. But existing aquatic juvenile stages have become independently different from their early ancestors – in some groups by the loss of spiracles and so on – and therefore cannot represent an ancestral stage. Nevertheless, amphibious life does not inevitably engender the evolution of distinctly aquatic structural adaptations. For instance, the thoracic spiracles of ephemeropteran nymphs have persisted through 250 million years of aquatic adaptations and, in the nymphs of Odonata, the abdominal spiracles are closed, while the thoracic spiracles open and function in the final instar before metamorphosis takes place. The possibility of an aquatic ancestry for pterygote insects cannot therefore be dismissed out of hand.

4.4 Tracheal Gill Theory

During the last century, the German entomologist C. Gegenbaur proposed a theory according to which insect wings were derived from the leaf-like tracheal gills of aquatic insects. In its general form, the character of its venation, the nature of its articulation and its constant vibratory movement, the tracheal gill of a mayfly nymph is very suggestive of a primitive wing. The fact that tracheal gills and gill-covers are today found entirely upon the abdomen does not preclude the possibility of their former development on the thorax and subsequent transformation into wings. The principal argument against this hypothesis in the past has been the fact that insects are primitively terrrestrial and not aquatic. Indeed, for many years the tracheal gill theory was dismissed with little consideration and, even today, has many opponents.

In contemporary entomology, the morphological characters of insects are not always treated according to their phylogenetic importance and fossil evidence often gives clues for quite different interpretations. For instance, all primitive Palaeozoic pterygote nymphs are now known to have had articulated, freely-movable wings reinforced by tubular veins. The significance of this has been discussed at length by Jarmila Kukalova-Peck (1978), who points out the implication that early pterygote wings were engaged in flapping movements. The immobilized, fixed, veinless wing-pads of recent nymphs have resulted, as already mentioned, from a later adaptation affecting only juveniles. If this is so, the paranotal theory of the origin of wings cannot be valid. The wings of Palaeozoic nymphs were curved backwards in Palaeoptera and flexed backwards at will in Neoptera. Consequently, the fixed oblique-backwards lobes of the wing pads in all modern nymphs must be secondary and is not homologous in Palaeoptera and Neoptera. Fossil evidence indicates that the major steps in evolution, which led to the origin, first of Pterygota, then of Neoptera and Endopterygota, were triggered by the origin and diversifications of the flight apparatus. Very probably the major events in pterygote evolution occurred first in the immature stages.

To summarize: for many years the paranotal lobe theory of the origin of wings has been widely accepted. Wings were derived from paranotal planes used for gliding, which then acquired muscles to control their inclination, and finally muscles which could move them up and down and thus lead to the evolution of flapping flight. More recently, however, there has been growing evidence to suggest that wings may have evolved among the nymphs of secondarily aquatic ancestral insects.

4.5 Selection for Flight

It is generally agreed that aerial dispersal is an important element in the lives of many species of insects, and wings are important in helping take-off and increasing buoyancy. This function would account for the tendency of wings to be confined to the sexually mature adult, the stage in which dispersal mostly occurs. Opinion is divided, however, as to whether insects acquired the capacity for flight initially because of its aid in dispersal, or whether there was a secondary development and the primary selective function was escape from predators. In the absence of fossil evidence, there is complete uncertainty as to whether the first of the flying insects were small (in which case dispersal would probably have been the primary function of wings) or large (in which case escape from enemies might have been their primary function).

There is some measure of agreement that, whatever the original function of flight, the development of wings arose from the need to control direction and attitude just before landing. When spiders are falling, they appear to stabilize their bodies in the air with their legs and it is possible that the limbs of early insects could also have been used to some extent to control attitude. It would certainly have been advantageous not to land upside down in damp environments or on water.

If the first pterygote insects were large, with well-developed mesothoracic and metathoracic paranotal lobes, the problem arises as to whether these could have evolved from gliding planes into flapping wings. The ancestral pterygote must have launched itself from a leaf or branch by either a jump or by a run. The latter seems to be the more probable for the following reason: the earliest pterygote used its paranotal lobes for gliding. The inclination of these could, presumably, by then be altered for increased efficiency. The muscles of a cockroach concerned with walking and flying contract in an almost identical sequence. This suggests that a common central nervous pattern of control could have served the needs of both running and flying.

The archetypal pterygote walked along with wings beating as it moved. When danger threatened, it would run away, off the edge of the leaf or branch, and continue to run with both legs and wings flapping. The fact that the wings of a thoracic segment of an insect beat in phase while the legs move in antiphase raises difficulties, but could be explained by postulating that two wings on different sides of the body would have been beating out of phase with the wings of the same segments. It may be worth noting also that study of the aerodynamic behaviour of various shapes suggests that larger insects of perhaps 1 cm in length with small legs would have been capable of the best gliding performances — provided that suitable attitudes could be maintained by use of the legs, small changes in body shape, and in the inclination of the paranotal lobes or whatever other structures may have served as precursors of wings.

If, on the other hand, the tracheal gill theory of the origin of wings is accepted, the thoracic branchial plates of an aquatic pterygote were pre-adapted to facilitate passive transport, by wind, of insects stranded by the drying out of their habitats. In their general form, these gill-plates would also have been pre-adapted for conversion into wings in the articulation of their bases and in their musculature. The climatic conditions of Devonian times would have been in conformity with this theory.

Despite the numerous conflicting hypotheses that have been proposed, it seems most probable that insect wings evolved once only, towards the end of the Devonian period. They were derived from pleural, articulated structures, homologous with ephemeropteran gill-plates, rather than paranotal lobes although, at the present time, neither theory can be proved or disproved and there is evidence both for and against each of them. If the archetypal pterygotes were large, the paranotal lobe theory would be more likely to be correct, and the initial function of flight would have been escape from enemies by jumping from trees, etc. If the first of the pterygotes were small, the tracheal gill theory would be the favourite, and the initial function of flight would have been dispersal from one moist habitat to another in a predominantly arid environment.

4.6 Phylogeny of the Lower Insect Orders

4.6.1 Fossil Evidence

Turning now to relationships between the various insect orders, we find a comparable degree of ignorance. The earliest fossil insect remains, found in Carboniferous rocks, are those of exopterygote types with moderately efficient wings. These fossils are accompanied by fragments of contemporaneous nymphs which indicate that the insects were undoubtedly hemimetabolous in their development. Some, which exhibit orthopteroid features, are grouped together as Protorthoptera. Others comprise an assemblage of primitive types and are included in the order Palaeodictyoptera. The Protorthoptera show a definite advance over the Palaeodictyoptera. Their thoracic structure suggests that they could run well, whereas the latter probably had feeble powers of movement on land. The Protorthoptera also show increased development of the hind wings, whose anal area is expanded. In contrast, the fore- and hindwings of the Palaeodictyoptera were very similar to one another.

The Protorthoptera are possibly connected with modern cockroaches through early roach-like 'Protoblattoidea', now regarded as primitive blattids rather than as a separate order. Other Protorthoptera resembled atavistic long-horned grasshoppers: some are primitive Orthoptera, others not. The Megasecoptera were more highly specialized than the Protorthoptera, and probably represent an entirely different line of descent from the Palaeodictyoptera. They are often grouped together in the Palaeoptera, an assemblage that includes the fore-runners of the Ephemeroptera and Odonata.

Yet another offshoot of the Palaeodictyoptera is the Paraplecoptera or Protoperlaria, which is clearly allied to existing Plecoptera. From the ancestral Paraplecoptera may also have evolved the Embioptera and Isoptera, but this is by no means certain. Certainly, termites must have evolved from cockroach-like forms, which later developed a complex social organisation, but there is, as yet, no palaeontological evidence for this. The Protohemiptera may also have arisen from a specialized family of Palaeodictyoptera. The venation of their wings was primitive, but they had sucking mouthparts like those of modern Hemiptera. How these can have evolved from the generalized biting jaws of the Palaeodictyoptera is not yet known, and this hypothesis is very unlikely. An extinct family of Lower Permian Homoptera, known as the Archescytinidae, has wing venation similar to that of early types of Psocoptera which have been found along with them. So the Palaeodictyoptera may have led via the Protohemiptera to the Archescytinidae, the Psocoptera, and the Siphunculata. The latter probably evolved as descendants of wingless psocids which became adapted for parasitism.

4.6.2 Palaeoptera and Neoptera

The Palaeodictyoptera is the oldest and most primitive order of insects known, and its wing venation is more complete than in any other insects. Its outstanding features include the greatly developed branching of most of the principal veins. Recent work has shown that all the Palaeodictyoptera, like the Megasecoptera, have highly modified piercing and sucking mouthparts. These two orders are now seen to be two most highly successful taxa of sap-sucking insects. Neither of them can be ancestral to any modern insect orders. The Palaeodictyoptera are, nevertheless, very primitive, and represent a pre-Palaeopteran offshoot from the insect stem. Research workers today attempt to trace the ancestry of individual monophyletic groups as far back as possible, and to associate the oldest known forms with their respective groups. Nevertheless, however consistently these aims are pursued, an impasse is eventually reached when the palaeontological record becomes so fragmentary that it is impossible to determine the diagnostic characters of the fossils.

Like the Palaeodictyoptera and the related Megasecoptera, the Odonata and Ephemeroptera are unable to fold their wings back over the abdomen. This is one of the main characters separating the primitive Palaeoptera from the Neoptera. The discovery, in 1952, that the Palaeodictyoptera had advanced haustellate mouthparts has shown that this order is not a primitive ancestor of all the Pterygota as had once been thought. On the contrary, both the Palaeodictyoptera and the Megasecoptera, which also had haustellate mouthparts, appear to represent a remarkable radiation of insects that fed either on liquids or on spores and pollen — both functions have been attributed to their peculiar mouthparts.

The wing venation of the Palaeodictyoptera, too, is in some respects less primitive than that of Permian Ephemeroptera. They retain a complete series of the main longitudinal veins, alternatively convex and concave (Sect. 4.7), but they usually lack intercalary veins, so that the distal areas, supported by the branched veins, are flat. The Palaeodictyoptera were part of a varied complex of orders showing a diversity of

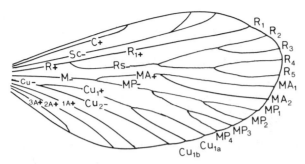

Fig. 21. Ancestral wing venation based on the work of J.H. Comstock and J.G. Needham. (After Imms 1931), *C* costa; *Sc* sub-costa; *R* radius; *Rs* radial sector; *M* media; *Cu* cubitus; *1A, 2A, 3A* anals

wing structures. Although most have been preserved with their wings stretched out, like those of the Adnisoptera, the Diaphanopteroidea, at least, had developed the ability to fold their wings over the abdomen. Outstretched wings do not permit mobility on the ground, and most palaeodictyopteroid insects were probably active flyers. The subject has been reviewed by R.J. Wootton (1976), who assumes a group to be monophyletic unless there is definite evidence to the contrary — but in the present state of knowledge, it is conceivable that either or both the Palaeoptera and the Neoptera may be polyphyletic.

The derivation of pterygote insects from apterygote Thysanura is now almost universally accepted. The important research of A. Lameere in the early 1920's suggested that the Pterygota should be divided into two groups — the Palaeoptera or Palaeoptelota, including the Ephemeroptera (mayflies) and Odonata (dragonflies), and the Neoptera, to which it has been estimated that 97 per cent of recent insect species belong. It is now generally believed that the pterygotes evolved along four different evolutionary lines: palaeopteroid, and neopteroid (orthopteroid, hemipteroid and neuropteroid), as we have seen.

4.7 Wing Venation

Fossil insects are often represented only by their wings. No doubt, soon after death, the bodies decay and disintegrate, while the light wings float away, to accumulate in certain geological strata where their durable structure enables them to resist destruction. Moreover, many insectivorous animals discard the wings. The result has been that insect palaeontology must inevitably depend mainly upon wing venation, since the structure of other parts of the body is little known, although more confirmatory evidence from this source is becoming available at the present time.

Determination of the origin and homologoies of insect veins was, for many years, complicated by the various systems of nomenclature employed by different investigators. However, in 1886, J. Redtenbacher published an important monograph in which he proposed a uniform system in which the principal veins were named costa, sub-costa, radius, media, cubital and anal respectively. Furthermore, he reaffirmed the fact

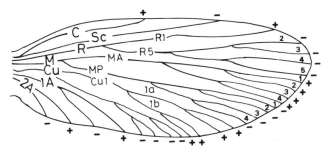

Fig. 22. Ancestral wing venation as envisaged by A. Lameere. (After Imms 1931), *MD* anterior branch of median; *MP* posterior branch of median. Other conventions as in Fig. 21

Fig. 23a,b. Branching of the cubital vein. a According to Comstock and Needham; b according to R.J Tillyard. (After Imms 1931). Conventions as in Fig. 21

that the main veins primitively alternate between convex and concave, and claimed that the earliest insects exhibit the richest venation: this is reduced in more recent forms. Even so, attempts to introduce a common terminology achieved little success until the work of J.H. Comstock and J.G. Needham, based on the homologies and development of the wing veins in all orders of insects, was published in 1898 and 1899. By an extensive study of the tracheae which precede, and later coincide with, the positions of the veins, these authors subsequently constructed an hypothetical ancestral system of venation from which all others might be derived (Fig. 21).

The original conceptions of Comstock and Needham have, however, subsequently been modified. Less emphasis is now placed on tracheation, while the important work of Lameere, published in 1922, afforded strong support to the earlier views of Redtenbacher, Lameere's study of the Palaeodictyoptera led him to stress the importance of following the alternation of convex and concave veins. He emphasized the fact that among the lower and more primitive orders of insects, the veins are alternately *convex* (+), when they follow ridges, and *concave* (−) when they lie in the furrows between them. The wing itself can thus be compared with a partly-opened fan. These features are clearly seen not only in the early fossil orders, but also in modern Palaeoptera — Odonata and Ephemeroptera — in which the alternation of convex and concave veins probably helps to increase the rigidity of the wings. Among higher orders, this fluting is obscured. The hypothetical ancestral scheme of wing venation, as envisaged by Lameere, is illustrated in Fig. 22.

In the hypothetical, primitive pattern of wing veins postulated by Comstock and Needham (Fig. 23), it can be seen that the costa (C) is unbranched and convex, while the subcosta (Sc) is forked and concave. The radius (R) divides into two, the R, which is convex and the radial sector (Rs) which is concave and, in turn, subdivides into two and then four veins (R_2-R_5). The media (M) divides into a convex two-branched anterior media (MA) and the concave, four-branched posterior media (MP). The

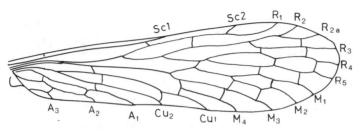

Fig. 24. Basic wing venation of the Panorpoid complex. Conventions as in Fig. 21

cubitus (Cu) again divides into two main branches, the first (Cu_1) being convex, the second (Cu_2) concave. The Cu_1 divides into an anterior (Cu_{1a}) and a posterior (Cu_{1b}) vein. Then follow three anal veins (1A to 3A) which are usually convex although 2A may be concave. The R_1 and Cu_1 veins are strongly convex in many insects and can therefore easily be recognized. This is helpful in identifying the other veins.

The division of the median vein by Lameere into a convex MA and a concave MP (Fig. 22) is of great importance because it implies that the latter is the counterpart of Comstock's media. This interpretation was, however, not advanced by Lameere, who believed that MP is absent in Endopterygota, being represented by a rudiment which had previously been regarded by R.J. Tillyard as M_5. Since M_5 is convex (+), however, Lameere's homology is improbable. It is, however, only fair to point out that, in his earlier writings, Tillyard had stated definitely that M_5 was concave (−), and Lameere was evidently misled by this.

In his work on the 'Panorpoid complex', described below, Tillyard concluded on ontological grounds that the cubital vein is primarily three-branched (Fig. 23b) and not two-branched as Comstock and Needham had believed (Fig. 23a). Thus the vein which the latter authors maintained to be the first anal (1A), which had secondarily become associated with the cubital system, was regarded by Tillyard as Cu_2. In other words, the primary bifurcation of the cubital vein takes place much nearer the base and the forked cubital vein of Comstock and Needham represents its anterior branch only. Tillyard's view is now generally accepted.

When Lameere's hypothetical scheme is applied to recent insects, several important modifications become apparent. The media is rarely represented by both MA and MP, although Tillyard considered this condition to be retained in the Ephemeroptera. Only MA is found in Odonata and Plecoptera. In all other orders the media (M) is represented by its posterior branch (MP). (For further details see Imms 1931).

4.8 The 'Panorpoid Complex'

Inter-relationships between most of the endopterygote orders are a matter for seeculation. At the present time, the origins of the Hymenoptera and Coleoptera are unknown; while the Siphonaptera (= Aphaniptera) or fleas have been variously derived from Diptera, staphylinid beetles, Trichoptera and the extinct Permochoristidae. In the remainder of this chapter, therefore, we shall consider only the Mecoptera, Neuroptera, Trichoptera, Lepidoptera, Diptera and certain related extinct orders

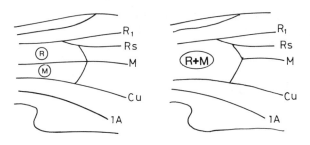

Fig. 25. Reduction in venation by atrophy. Conventions as in Fig. 21

Coalescence

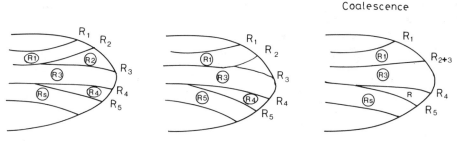

Fig. 26. Reduction in venation by coalescence. Conventions as in Fig. 21

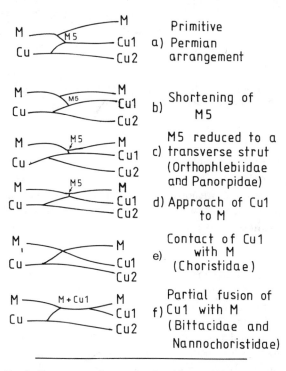

a) Primitive Permian arrangement

b) Shortening of M5

c) M5 reduced to a transverse strut (Orthophlebiidae and Panorpidae)

d) Approach of Cu1 to M

e) Contact of Cu1 with M (Choristidae)

f) Partial fusion of Cu1 with M (Bittacidae and Nannochoristidae)

Evolutionary changes in the cubito-median Y-veins of the forewing of the Mecoptera.

Fig. 27a–f. Evolutionary changes in the cubito-median Y-vein of the forewing of the Mecoptera. a Primitive Permian arrangement; b shortening of M5; c M5 reduced to a transverse strut (Orthophleobiidae and Panorpidae); f partial fusion of Cu1 with M (Brittacidae and Nannochoristidae). Conventions as in Fig. 21

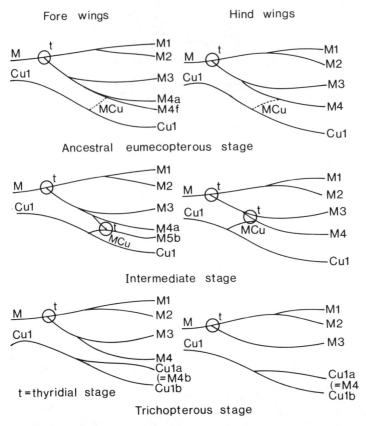

Fig. 28. Evolution of a trichopterous from a eumecopterous type of venation. *Left* forewings; *right* hindwings; *top* ancestral eumecopterous stage; *centre* intermediate stage; *bottom* trichopterous stage. Conventions as in Fig. 21

which, in the opinion of most palaeontologists and entomologists, form a single mono-phyletic group, the 'Panorpoid complex', or Mecopteroidea. From this, only the Hymenoptera, Coleoptera and their parasitic offshoot, the Strepsiptera, are excluded. These present unsolved phylogenetic problems, to which palaeontology has contributed relatively little. It is not impossible, however, that the Hymenoptera may be related to the Eumecoptera, in which case this group would also form a part of the Panorpoid complex.

The basic wing venation of the Panorpoid complex is shown in Fig. 24. Most of the principal veins and their primary branches are represented in *Panorpa*'s archaic pattern of venation. Reduction in venation can take place in two ways: (a) by atrophy (Fig. 25), and (b) by coalescence (Fig. 26). Figure 27 illustrates evolutionary changes in the cubito-median Y vein of the forewing in Mecoptera and, in Fig. 28, the evolution of a trichopterous from a eumecopterous type of venation is shown. All primitive Trichoptera differ from the Eumecoptera in the following: (a) Cu_1 is not simple but strongly forked, (b) M has four branches in the forewing and three in the hindwing, (c) the

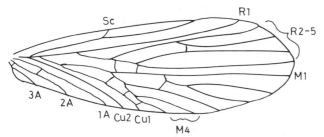

Fig. 29. Wing of *Permochorista jucunda* (Lower Triassic). (Adapted from R.J. Tillyard). Conventions as in Fig. 21

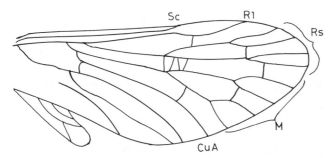

Fig. 30. Wing of *Pseudodiptera gallica* (Lower Triassic). (Adapted from R.J. Tillyard). Conventions as in Fig. 21

anal veins of the forewing are linked in a double Y-vein (Fig. 27). These characters might seem to preclude derivation of the Trichoptera from the Eumecoptera, in which Cu_1 is universally simple and the branches of M are more numerous. In the Permian Eumecoptera, however, there is a reduction of M in the hindwing compared with the forewing, which suggests that the Trichoptera may have arisen from a primitive eumecopterous type in which the hindwing already had one branch of M less than the forewing. Perhaps the deep fork of Cu_1 arose, at one stroke, by coalescence with the posterior branch of M.

In a comparable manner, the Diptera could be derived from the Permochoristidae (Fig. 29) and possibly the Paratrichoptera by simplification and reduction of the venation (Figs. 30–33); and the ditrysian Lepidoptera via the Trichoptera and Heteroneura (Fig. 34). Tillyard's basic conclusions are summarized in Fig. 35, which has been updated to take account of later opinions.

The early Protomecoptera had profusely branched venation and included forms that resembled the relict genus *Merope*. They might possibly have been ancestral to the Neuroptera. The Paramecoptera of the Upper Permian were an offshoot from the Permochoristidae, and thus allied to the ancestral Trichoptera and Lepidoptera, linking them with the Mecoptera. Confirmation of the affinities of the Trichoptera and Lepidoptera is also afforded by structural characteristics, many of which also apply to Diptera. The Triassic Paratrichoptera (= Mesopsychidae) may have been ancestral to the Diptera but, since the Nannochoristidae (Mecoptera) have a dipterous-

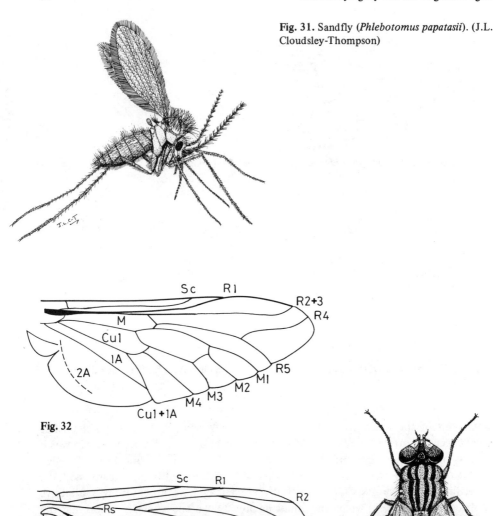

Fig. 31. Sandfly (*Phlebotomus papatasii*). (J.L. Cloudsley-Thompson)

Fig. 32

Fig. 34   **Fig. 33**

Fig. 32. Wing of housefly (*Tabanus*). Conventions as in Fig. 21

Fig. 33. Housefly (*Musca domestica*). (J.L. Cloudsley-Thompson)

Fig. 34. Hepialid wing (Lepidoptera: Homoneura). Convention as in Fig. 21

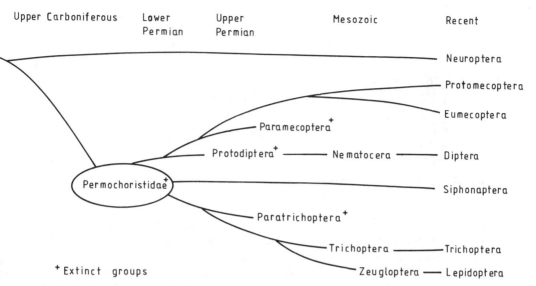

Fig. 35. J. Tillyard's conclusions (updated) regarding the 'Panorpoid complex'

like venation – more complex than that of the Panorpidae – it is more probable that both Diptera and Nannochiristidae are descended directly from the Permochoristidae.

4.9 Insect Flight

The success of insects, both in numbers of species and of individuals, can be attributed to several adaptive features, one of which is the power of flight. The value of wings is indicated by the fact that there are about 100 times more winged species of insects than there are of wingless species. Yet, the most successful order of all – the Coleoptera – is an order containing an unusually large number of species that have become secondarily wingless!

The wings of primitive, ancestral insects were held permanently extended on each side of the body, as they still are in the Anisoptera or held vertically as in Zygoptera and Ephemeroptera (Sect. 4.6.2). Such wings are moved by muscles attached directly to the base of the wing, but, in most orders of modern insects, two changes have taken place: (a) Muscles have evolved which fold the wings backwards over the abdomen; (b) the wings are powered by indirect wing muscles attached to the walls of the thorax, and other muscles control the inclination of the wing (Fig. 36). Direct wing muscles are found in Blattaria as well as in Odonata, while, in Orthoptera and Coleoptera, the downward movement of the wings is produced by the direct and indirect muscles acting together. There are two pairs of indirect wing muscles: these operate by distorting the shape of the thorax. Contraction of the inner horizontal muscles depresses the wings, while contraction of the outer, vertical muscles causes them to rise. Relatively little work is actually performed in raising and lowering the wings:

Fig. 36. Operation of direct and indirect flight muscles in insects

most of the effort is spent in distorting the shape of the thorax. Since the thoracic walls are elastic, however, much of this energy is stored and used to assist the opposing muscles during the next stroke.

High rates of wing beat are achieved by means of a 'click' mechanism. When the wings are being moved up or down, their movement is at first resisted by the elasticity of the thorax. After a certain point, however, this resistance vanishes and the wings click suddenly into the new position. Contracting muscles relax as soon as the click point has been reached, while the opposing muscles which are suddenly stretched, immediately begin to contract. There is thus an oscillating system which can operate at almost any speed, depending upon the elastic structure of the thorax.

The rate of wing beat varies from about 8–12 per second in butterflies, 20–30 in dragonflies, and 200 in houseflies, to 600 in mosquitoes, and up to 1000 or more beats per second in the ceratopogonid midge *Forcipomyia*. For many years, it was a puzzle how an insect could beat its wings so fast, because nerves cannot convey a succession of stimuli to muscles and cause them to contract and relax in such quick succession, in view of the 'refractory period' which limits the frequency at which impulses can follow one another. In the presence of an action potential, no further stimulation can initiate another impulse, nor can an impulse from elsewhere pass through the area. This effect lasts for 2 or 3 ms. The explanation of high rates of wing beat lies in the click mechanism, described above and, secondly, in the fact that the articulation at the base of the wing automatically changes its inclination as the wing moves up and down. The nerves to the wing muscles need only set the flight mechanism in motion and keep the muscles in such a tonic state that they keep on contracting spontaneously in response to sudden stretching.

I shall not discuss the mechanism and control of insect flight in detail, because it has been treated fully in many standard entomological text books. Instead, I will merely explain a couple of simple points which help in understanding the principles involved. In the first place, the forces acting on the wings of an insect, like the aerodynamic forces on the wings of an aeroplane, can be resolved into two components, lift and thrust. To keep an insect steady in the air, lift must roughly equal the insect's weight while thrust is the horizontal force that drives it forwards. When the insect is flying at constant speed, this compensates for the drag force. It is impossible to produce lift without drag, but the latter can be reduced by streamlining the body and varying the shape of the wing and its inclination. Higher ratios of lift to drag are obtained with longer, narrow wings than with short, broad ones; but excessively long wings would be disadvantageous because their bases would have to be extremely heavy to be strong enough. Lift is produced by deflecting air downwards, and by the suction effect of air passing over the upper surface of the inclined wing: it acts at right angles to the direction of motion. The effect of this is demonstrated every time two sheets of paper adhere to one another and are separated by blowing air over their surfaces!

A four-winged insect is inherently stable since its centre of gravity falls between the forces of lift engendered by the mesothoracic and metathoracic wings. Similarly, and by analogy, a car is stable because its weight falls within the area supported by the four wheels. In dragonflies, the legs are thrust forward by a distortion of the thorax so that they can be used as a trap to catch prey in flight. In consequence, the wings are rotated into a posterior direction so that lift would be exerted behind the centre of gravity, were this not moved backwards too by the elongation of the abdomen.

If the number of wings, or wheels, is reduced to two, balancing becomes necessary. Diptera, which have lost their second pair of wings, require even more sophisticated flight control than do four-winged insects. This control is provided by an air-speed indicator and a gyroscope. The long antennae of Nematocera, and the antennal aristas of Brachycera and Cyclorrhapha are deflected by the airstream in proportion to the speed at which the insect is flying. This deflection stimulates a chordotonal sensillum (Johnston's organ) lying in the second segment of the antenna, with its distal insertion in the articulation between the second and third segments. Johnston's organ occurs in all adult insects but not in Diplura or Collembola. It is concerned with the maintenance and control of flight speed. It is also a sound receptor, a gravity receptor and, in aquatic forms, registers water movements or ripples.

The metathoracic wings have been modified in Diptera to form 'halteres': in male Strepsiptera, the forewings form similar dumb-bell-shaped structures. The halteres of flies vibrate with the same frequency as do the forewings, but in antiphase, and act like alternating gyroscopes. As in a gyroscope, the moving halteres possess inertia, and any tendency to change their direction is resisted, affecting campaniform sensilla in the cuticle near their bases. Stimulation of these by yawing leads to a reflex modification of the twisting of the forewings during the downstroke, so that the deviation is corrected. Rolling and pitching also produce torque at the bases of the halteres at right angles to the stroke plane, but these various torques are differentiated by the timing and frequency of their operation.

Further Reading

Bristowe WS (1958) The world of spiders (The New Naturalist). Collins, London
Comstock JH (1918) The wings of insects. Comstock, New York
Henning W (1981) Insect phylogeny. John Wiley, Chichester New York
Hinton HE (1953) The phylogeny of the panorpoid orders. Ann Rev Entomol 3:181–206
Imms AD (1931) Recent advances in entomology. Churchill, London
Kukalova-Peck J (1978) Origin and evolution of insect wings and their relation to metamorphosis, as documented by the fossil record. J Morphol 156:53–125

Kukalova-Peck J (1983) Origin of the insect wing and wing articulation from the arthropod leg. Can J Zool 61:1618–1668

Manton SM (1964) Mandibular mechanisms and the evolution of arthropods. Phil Trans R Soc (B) 247:1–183

Manton SM (1977) The Arthropoda. Habits, functional morphology and evolution. Clarendon Press, Oxford

Richards OW, Davies RG (1977) Imm's general textbook of entomology (10th edn), vol 1. Structure, physiology and development. Chapman and Hall, London; John Wiley, New York

Tiegs OW, Manton SM (1958) The evolution of the Arthropoda. Biol Rev 33:255–337

Wigglesworth VB (1976) The evolution of insect flight. In: Rainey RC (ed) Insect flight. Symp R Entomol Soc Lond 7 (12):255–269

Wootton RJ (1976) The fossil record and insect flight. In: Rainey RC (ed) Insect flight. Symp R Entomol Soc Lond 7 (12):235–254

Wootton RJ (1981) Palaeozoic insects. Ann Rev Ent 26:319–344

5   Evolutionary Trends in Reproduction

5.1 Spermatophores and Their Phylogenetic Significance

Arthropods evolved in the sea, from the same ancestral stock as that from which the annelids arose. Reproduction can take place in several different ways in aquatic environments. Many marine taxa, including coelenterates, molluscs and echinoderms, merely shed their sexual products into the water, where the eggs are fertilized. Although this method is extremely simple, nevertheless the sexes must be in fairly close proximity to ensure union of egg and sperm, whose vitality is of limited duration. Furthermore, the reproductive activity of the sexes must be synchronized. It is usually correlated with the tides and phases of the moon. Efficiency is improved when the sexes come together, as in *Limulus,* the male fertilizing the eggs as soon as they have been laid. Even so, this could not take place on land without eggs, sperm, and the resulting zygotes and embryos suffering from desiccation.

Wasteful dilution of eggs and sperm can be avoided in one of two ways: either by internal insemination or by the formation of spermatophores. These are small packets of sperm which are transferred from the male to the female. Spermatophore production is found among leeches, cephalopods and newts, for example. Internal insemination may be achieved by means of gonopodial appendages, or through true copulatory organs. Modern annelids not only discharge their eggs and sperms into the sea, but some of them produce spermatophores, while others indulge in copulation. Presumably the terrestrial arthropods are descended from annelid ancestors that transferred spermatophores, or else had copulatory organs for the insemination of free sperm, since these are the pre-adaptations necessary for the transition from aquatic to terrestrial life. As we shall see, the first of these possibilities appears to have been the case. The first of the terrestrial arthropods probably transferred spermatophores from males to females.

Many different modes of sperm transfer have evolved among crustaceans. Several copulate, using modified limbs as copulatory organs and as a means of sperm transfer, while spermatophores are also often formed and transferred, more or less directly, into the female genital openings. Crustacean spermatophores, however, are primitive. They lack any specific covering or supportive elements, especially in those Decapoda that have developed gelatinous sperm balls.

In contrast, crabs and woodlice, which have successfully conquered the land, have gonopods that are well adapted for transferring sperm packets directly into the genital openings of the females. The great variety of structures and methods used for sperm transfer among crustaceans suggests that these are adaptive rather than phylogenetic features. The study of reproductive biology cannot therefore contribute much to elucidating systematic relationships within the class. A comparable situation exists among the terrestrial arthropod taxa. The modes of sperm transfer are so variable that no major phylogenetic conclusions can be drawn from their study, although some indications may be obtained by this means of the relationships between different orders within the same class.

Insemination in terrestrial Arthropoda takes place primitively by means of spermatophores which are transferred indirectly. From this basic trait, evolution appears to have followed one of two paths — either to copulation which may, in turn, lead to loss of the spermatophore and the transfer of free sperm, or to indirect free sperm transfer. Pairing is common in most groups, and this is not infrequently preceded by aggregation and/or courtship.

5.2 Functions of Aggregation and Courtship

Among Collembola, Pseudosorpiones and oribatid mites, whose population densities are often extremely high, there is frequently no attraction of one individual to another. Males deposit spermatophores on the ground, apparently without the benefit of stimulation from the presence of females, and these are later taken up by passing females who may never encounter a male. (This condition must necessarily have arisen secondarily from the direct transfer of spermatophores from the male to the female). More often, however, mating is preceded by aggregation or courtship.

5.2.1 Aggregation and Swarming in Insects

Insect swarms may, or may not, be a response to stimuli provided by other members of the same species. In the case of Diptera Nematocera, Stratiomyidae, Tabanidae and so on, aerial swarms take up more or less fixed positions in relation to stationary objects or markers. For instance, *Culicoides nubeculosus* swarms over dark patches of damp sand, or above pats of cow dung, *Serromyia* spp. beneath the tips of branches silhouetted against the sky. The lesser housefly, *Fannia canicularis,* also seems to use overhead markers. The timing of crepuscular swarms appears to be related to light intensity.

The function of swarming is by no means clear because fly swarms usually, although not invariably, consist entirely or predominantly of males. They are, however, conspicuous to females, which sometimes visit them and find mates. Many modern observers, nevertheless, have doubted whether any significant proportion of females finds a mate in this way. They have conceded that swarming may have evolved as a mating device, but that it represents a habit which has persisted long after its original function has disappeared. (If so, this habit cannot render the insects that indulge in it more vulnerable to predators than would otherwise be the case, or it would have been selected against). Alternatively, swarming may provide the psychological stimulation that the males need before mating takes place: but why this stimulation should be necessary in some species and not in others is, again, not clear.

Experiments on the chironomid midge *Spaniotoma minima* indicate a more probable explanation. Unmated males and females, newly released, establish swarms in which both sexes participate. Matings soon take place, however, and the mated females do not rejoin the swarm. Consequently, the sex ratio of the swarm changes rapidly from one of equality to a preponderance of males. The failure of mated females to return to swarms has also been recorded in species of *Chironomus, Trichocera* and *Dixa*. Aquatic insects such as Ephemeroptera, Trichoptera and Plecoptera swarm over water. Females fly into these swarms and mating takes place.

5.2.2 Courtship

Interaction between male and female arthropods prior to mating may be as brief as a few seconds, or may last for several hours spread over a number of days. Brevity of interaction is not, however, indicative of simplicity in courtship: even during the brief contacts between male and female houseflies, for example, the male may provide the female with a number of chemical, tactile and auditory cues.

The functions of courtship in arthropods are numerous and diverse. In many cases they are not properly understood. One of them is the initial attraction of a mate, sometimes over a considerable distance. Many moths have evolved wind-borne pheromones which enable males to find the females that produce them. For instance, the male silk moth *Bombyx mori* is able to fly up-wind to females several kilometres away. Scent production by female insects frequently results in large aggregations of males. Day-active species often employ visual displays, while nocturnal forms tend to rely on auditory or olfactory stimuli. Bush crickets or long-horned grasshoppers (Tettigoniidae) and crickets (Gryllidae) stridulate, while male fireflies (Lampyridae) attract females by producing flashes of light as they fly through the air.

The potential information conveyed by courtship among insects includes indication about oviposition sites, potential male parental care, the genetic quality of the male, and mate discrimination by males. The subject has been discussed in considerable detail by Thornhill and Alcock (1983). Courtship may serve to stimulate the female and release mating behaviour. It can also serve to block hunger drives in predatory species where the female is larger and more powerful than the male.

The suggestion has been made that, in the case of spiders, courtship may enable the female to recognize the male, thus suppressing her predatory instincts, and that it also stimulates her. If, however, it serves as a releaser, there is no need to postulate recognition. Jumping-spiders (Salticidae) which hunt their prey by sight, indulge in visual courtship displays. For instance, the male *Hasarius adansoni* extends his forelegs, with their conspicuous white tarsi, and advances slowly, waving them up and down (Fig. 37). Occasionally, he leaps to one side or other of the female, and then continues to approach. *Marpissa muscosa* males likewise raise their forelegs in the presence of females, but somewhat higher. The abdomen, too, is lifted steeply upwards and the spider zigzags rapidly from side to side as he approaches the female (Fig. 38). Display in jumping-spiders cannot be released by touch, although this plays an important role in the final stages of courtship. Neither is display released through chemical stimulation alone although, in the presence of certain visual stimuli, these are important secondary releasers of courtship.

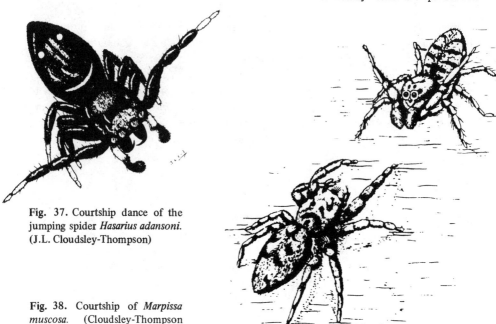

Fig. 37. Courtship dance of the jumping spider *Hasarius adansoni*. (J.L. Cloudsley-Thompson)

Fig. 38. Courtship of *Marpissa muscosa*. (Cloudsley-Thompson 1976 after W.S. Bristowe)

The courtship display of wolf-spiders (Lycosidae and Pisauridae) is also visual but to a lesser extent. In most males of the genus *Pardosa* the palps are conspicuously black. They play the principal part in courtship poses, their semaphore signalling being supplemented by movements of the legs and vibrations of the abdomen.

In short-sighted, nocturnal hunting spiders, which detect their prey by the sense of touch, courtship is primarily tactile. In *Dysdera,* the forelegs of both male and female are held aloft, quivering, and each spider strokes the other intermittently. Finally, in the case of web-spinning spiders which react to the vibrations caused by the struggling of an insect in their snare, the courtship of the male consists in tapping a distinct code on the web of the female, as a result of which he is not mistaken for prey.

Various hypotheses have been proposed to explain the phylogeny of courtship. According to these, spider display has evolved from motions of self-defence, is a manifestation of the excitement of the male, or results from chemotactic searching movements. Most biologists agree with the last interpretation. Courtship occurs very sporadically among insects. Some display only the basic minimum necessary for sexual identification. In others, complex courtship behaviour takes place before mating. Male courtship consists of a sequence of activities which varies in different species of *Drosophila*. In *D. melanogaster* (Fig. 39) six elements have been identified as follows: (a) *Tapping.* The male approaches the female and taps her with his forelegs. (b) *Orientation.* He then faces the female and follows her if she moves. (c) *Scissoring.* The male displays with a scissoring movement of his wings. (This element occurs infrequently in *Drosophila melanogaster*). (d) *Vibrating.* One wing is extended and vibrated in the vertical plane. (e) *Licking.* The proboscis is extended and licks the female's genital

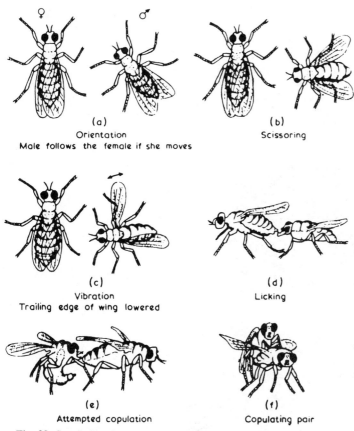

Fig. 39. Successive elements in the courtship display of the male fruit-fly *Drosophila melanogaster.* (Cloudsley-Thompson 1976 after A. Manning)

region. (f) *Attempted copulation.* The male curls his abdomen down and forward, and attempts to mount the female.

A receptive female spreads her genital plates and her wings which facilitates mounting, but an unreceptive female shows a variety of repelling movements, flicking the wings, kicking backwards with the hindlegs, extending the ovipositor, and jumping or moving rapidly away. Courtship in other species of *Drosophila* differs in the order in which the various elements appear, or the speeds at which they are performed.

The visual components of the male displays of *Drosophila* spp. are thought not to be means by which a female recognizes a mate of her own species. Recognition usually occurs at the onset of courtship when the male taps the female with his forelegs. The function of courtship may be that it enables a female to select a particularly virile mate and to reject poorer specimens.

Courtship may assist physiological maturation of the gametes (e.g. in the desert locust *Schistocerca gregaria*). It may also provide behavioral barriers to mating between different insect populations, thus preventing hybridization. No doubt its primary functions are appeasment and synchrony. In territorial insects, such as crickets

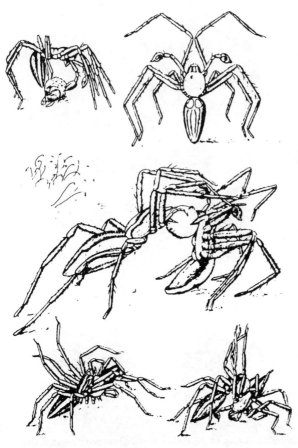

Fig. 40. The male wolf-spider *Pisaura mirabilis* wraps a fly, presents it to, and is accepted by the female. (Cloudsley-Thompson 1976 after W.S. Bristowe)

and Odonata, it may enable a female to enter the territory of a male without being attacked or driven off like any other intruder. In the dragonfly genera *Calopteryx* and *Pachydiplax,* sexual dimorphism has evolved and females are apparently recognized by sight, although there is little evidence of courtship display.

Territorial behaviour may have originated from sexual behaviour in cases where sexual dimorphism was initially slight, sexual discrimination poor, and homosexual approaches were, in fact, attempts to take up the tandem position until repulsed by other males.

Insects of many species tend to avoid one another and, in a wider sense, courtship serves to block the escape of the female. In the courtship flight of many butterflies the male precedes the female and prevents her from heading off. Again, just as the male wolf-spider *Pisaura mirabilis* gives a fly to the female as part of his courtship repertoire (Fig. 40), so female robber-flies (Empidae) are presented with prey by their suitors as part of an elaborate courtship. Copulation takes place while the female is engaged in eating the food thus provided.

In *Hilara,* the male fly binds the prey with silk secreted by his front tarsi. Other species bind up inedible objects such as petals, daisy florets, or particles of frass.

These serve temporarily to distract the attention of the female while mating takes place. The male scorpion-fly *Panorpa communis* (Mecoptera) secretes a drop of saliva onto the surface of a leaf: the female feeds on this while the male is fertilizing her. Since courtship feeding also occurs in non-predatory insects, its function is probably concerned more with diverting the female than with allaying her cannibalistic proclivities.

Two general patterns are recognized in the mating behaviour in cockroaches. Among species which adopt the first, exemplified by *Blatella germanica,* the male must make antennal contact with the female before courtship is initiated. Recognition appears to be mediated through contact chemoreception. In those of the second type, illustrated by *Periplaneta americana,* the female produces a volatile sex pheromone which attracts males from a distance and acts as the principal releaser of their courting behaviour. In *Byrsotria fumigata* a volatile pheromone produced by virgin females is the primary releaser of the male's response: he turns away from the female and opens his wings. The female then moves forward, straddling the male's abdomen while the latter, receiving tactile stimulation at the tip of his abdomen, moves backward. The female then makes feeding motions with her mouth parts over the dorsal surface of the male's abdomen. Meanwhile, genital connection is effected.

Courtship is frequently absent from the mating activities of arthropods. When it exists, however, it may be surprisingly complex. The evolutionary origins of courtship, from conflicts in the mating situation, can readily be discerned in vertebrates: in arthropods they are obscure. From the examples given above, it will be seen that their function may be to manoeuvre 'coy' females for copulation and prevent their escape; to facilitate the meeting of solitary animals, and to prevent cross-mating. Synchrony of physiological timing is less significant in Arthropoda than among vertebrates, and there is usually less need for appeasement. Courting males in communal groups may suffer from interference, but they usually repel other males rather than lure females away. Sometimes there are communal gatherings of males for communal advertisement, as when fireflies flash in unison and crickets sing in chorus. Perhaps the most important function of arthropod displays is to direct the mates' attention towards the relevant releasing stimuli.

5.3 Indirect Spermatophore Transfer via the Substrate

Indirect transfer of spematophores, via the substratum, must have evolved in Arthropoda soon after the conquest of the land. It is found in scorpions which are among the most atavistic of living Arachnida, in Chilopoda, Collembola and in various apterygote insects. Stalked sperm packets seem to have evolved secondarily in some water-mites (Hydracarina) (Fig. 41). It is not possible, in the case of arthropods which exhibit indirect spermatophore transfer, to determine when courtship ends and mating begins, since the two merge with one another.

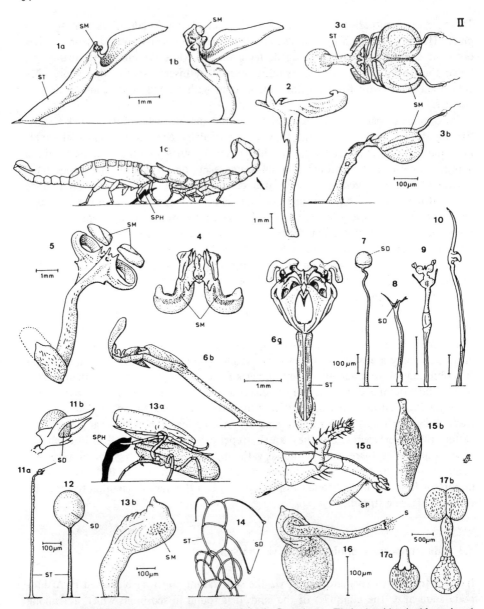

Fig. 41, 1–17. Spermatophores and sperm transfer in Scorpiones, Thelyphonida, Amblypygi and Acari. (Schaller 1979). Scorpiones: *1, 2: 1 Euscorpius italicus; 1a* spermatophore (lateral view); *1b* spermatophore (opened); *1c* pairing. (After H. Angermann from F. Schaller). *2* Spermatophore of *Heterometrus petersii* (lateral view). (H. Nemenz and J. Gruber). Pedipalpi: *3–6: 3* Spermatophore of *Trithyreus sturmi; 3a* from above; *3b* lateral view. (H. Sturn). *4* Spermatophore of *Mastigoproctus brasilianus* from above. (P. Weygoldt). *5* Spermatophore of *Damon variegatus* with detached sperm masses. (A.J. Alexander). *6* Spermatophore of *Tarantula marginemaculata; 6a* frontal view; *6b* lateral view. (P. Weygoldt). Acari: *7–17: 7* Spermatophore of *Belba gamiculosa*. (After F. Pauly from F. Schaller). *8* Spermatophore of *Cytalatirostris* lateral view. (G. Alberti). *9* Spermatophore of *Biscirus silvaticus* frontal view. (G. Alberti). *10* Spermatophore of *Bdella longicornis* lateral view. (G. Alberti). *11* Spermatophore of *Anystis baccarum*. (R. Schuster and

5.3.1 Arachnida

There are no known transitional forms in the series from aquatic to terrestrial life among Chelicerata, as illustrated by Xiphosura, Eurypterida, Pycnogonida and Arachnida. Pairing occurs in most, with the exception of some pseudoscorpions and mites (Sect. 5.2). Somewhat similar stalked spermatophores are found in scorpions, pedipalps (Thelyphonida and Phrynichida), and false-scorpions (Fig. 41). They are attached to the ground and taken up by the females after lengthy and complicated courtship.

The courtship dance of scorpions was first described in 1810, but spermatophore production and mating were not observed until 145 years later. In 1907, J.H. Fabre published a detailed and classic account of the courtship of *Buthus occitanus* in southern France, and claimed that it lasted for many hours. During the mating dance, the male scorpion uses his sensory pectines to detect a piece of stone or rock to which a spermatophore can be attached, which may take less than a minute. Initial contact between the sexes appears to be almost accidental but, immediately he has touched the female, the male scorpion shifts the grip of his pedipalps until he is clasping those of his partner. From time to time he vibrates his pedipalps and body, a movement that has been described as 'juddering'. The two scorpions now move back and forward in a 'promenade-à-deux' (Fig. 42) during which the pectines of the male are spread out and swept over the substrate. Occasionally the male moves forward and 'kisses' the female with his chelicerae. As soon as a spermatophore has been deposited, the male moves backwards again, pulling his mate over it. Orienting her body so that the paired sperm containers are situated exactly in front of her genital opening, she opens the cover of the spermatophore by a sudden movement of her body and the sperms are injected into her genital atrium. The male immediately releases the female and, in some species, she subsequently consumes the empty spermatophore.

Indirect spermatophore transfer in pseudoscorpions has long been known, and has been analyzed in detail by Weygoldt (1969), who distinguishes three types of behaviour as follows:

a) Without mating, the males and females acting completely independently in time and space. This type has already been mentioned (Sect. 5.2). Spermatophores are probably recognized through the presence of a pheromone. It is primitive, wasteful, and effective only in a humid environment and among gregarious species.

b) Without mating, but with the formation of spermatophores only in the presence of females. When a male encounters a female, he may grasp one or both of her pedipalps and then move away. Only after this does he deposit a spermatophore,

---

I.J. Schuster). *12* Spermatophore of *Calyptostoma velutinus*. (After G. Theis and R. Schuster). *13a* Pairing; *13b* spermatophore of *Saxidromus delamarei*. (Y. Coineau). *14* Combined spermatophores of *Nanorchestes amphibius*. (R. Schuster and I.J. Schuster). *15 Haemogamasus hirsutus; 15a* spermatophore grasped by chelicera of a male; *15b* spermatophore. *16* Spermatophore of *Uroobovella marginata*. (H. Faasch). *17* Spermatophore of *Ornithodorus savignyi, 17a* at beginning; *17b* at end of evagination. B. Feldman-Muhsam et al.; *S* Sperm; *SD* sperm drop; *SM* sperm mass; *SPH* spermatophore (sperm package); *SS* sperm sac; *ST* spermatophore stalk; *LD* liquid drop (with attractive pheromone?)

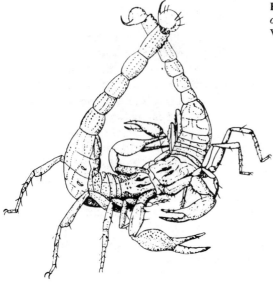

Fig. 42. Courtship dance of scorpions, *Buthus occitanus*. (J.L. Cloudsley-Thompson after M. Vachon)

whether the female is receptive or not. This type of sperm transfer is somewhat less inefficient than the one described above.

c) Sperm transfer with mating. Males and females act together in time and space, and the formation of a spermatophore is triggered only when a receptive female is present and prepared to accept it. In this way, the wastage of sperms is almost eliminated. In some families, the male grasps one or both palpal hands of the female in a manner reminiscent of the courtship of scorpions. In others, there is a retreat from bodily contact. The male courts the female by means of a series of body movements and display of the ram's horn organ. After deposition of the spermatophore, contact between the partners is established and the male assists the female in taking up the spermatophore (Fig. 43).

The reproductive biology of Thelyphonida (= Uropygi) is superficially similar to that of scorpions and false scorpions. The male touches the female with vibrating movements of his legs and grasps the antenniform front legs of the female with his palpal chelae. He then transfers these to his chelicerae and walks backward and forward for some hours, pulling and pushing the female as he does so. Eventually he releases her antenniform legs with his pedipalps, although continuing to hold them with his chelicerae, and turns so that both animals are facing in the same direction. The female now embraces the male's opisthosoma with her pedipalps. The male then deposits a spermatophore and pulls the female over it. Her gonopore opens and her genital operculum grasps the hooks of the sperm carriers and pulls them out of the spermatophore.

Pairing in Schizomida follows a phase of courtship in which the male chases the female, touching her with vibratory movements of his antenniform forelegs. She then turns round and he does the same, so that she now stands behind him. In this position, she hooks her chelicerae into the base of his peculiar knob-shaped metasoma and

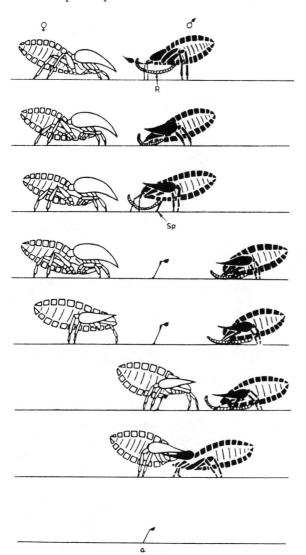

Fig. 43. Mating in the false scorpion *Chelifer concroides R* ram's horn organ of male; *Sp* spermatophore; *a* rod of spermatophore left behind after mating dance. (Cloudsley-Thompson 1976 after M. Vachon)

allows herself to be carried forward. The male deposits a spermatophore which contains two sperm balls. Repeatedly jerking his posterior, he finally pulls the female to the spermatophore and she takes up the sperm with her vulva.

The partners do not make firm bodily contact at any stage of reproduction in the Phrynichida (= Amblypygi). Thus, this order shows a tendency towards dissociation of the sexes as found in certain false scorpions, Collembola, Symphyla and other primitive mandibulates. The sex partners perform long palpating movements before the male finally turns round and deposits a spermatophore while maintaining contact with the female by means of his posteriorly-directed first pair of legs. After this, he turns again and signals her by vibrating his tactile forelegs. He then moves slowly backward, leading her exactly to the spermatophore. Pairing and indirect sperm transfer in other

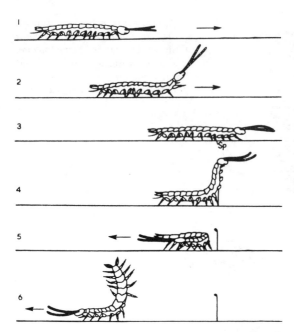

Fig. 44. Male symphylid depositing sperm stalks. (Cloudsley-Thompson 1976 after L. Juberthie-Jupeau)

species of Phrynichida proceed essentially in the same way, although the male may play a more active role in sperm transfer. He performs before the female intensive movements which resemble mock attacks. Then he turns round and deposits a spermatophore which he rubs with the anterior part of his body. The female then moves above the spermatophore, making repeated circulatory movements until the male seizes her with his pedipalps and drags her over it. She tears out the sperm packets with the claw-shaped sclerites of her genital operculum and moves away. Afterwards she usually eats the empty spermatophore.

The fact that both sexes face in the same direction in Thelyphonida, Schizomida and Phrynichida suggests a common ancestry for this group of arachnids, as compared with scorpions and pseudoscorpions in which the animals face each other during spermatophore production. The morphology of the spermatophores themselves likewise emphasizes the similarities between Thelyphonida and Phrynichida. In Scorpiones and Pseudoscorpiones, each spermatophore contains one single sperm mass enclosed in a sperm package which may have one or two openings. In the Thelyphonida, Schizomida and Phrynichida, on the other hand, the spermatophores carry two clearly distinct sperm masses, enclosed separately in two sperm packages or carriers.

Although most species of Acari reproduce by direct sperm transfer, some oribatid mites, like Collembola and certain pseudoscorpions, maintain high population densities and large numbers of spermatophores are deposited in the absence of females (Sect. 5.2). In other families, stalked spermatophores are transferred via the substrate after courtship displays (Fig. 42). This has almost certainly evolved independently of the same behaviour in other arachnid groups.

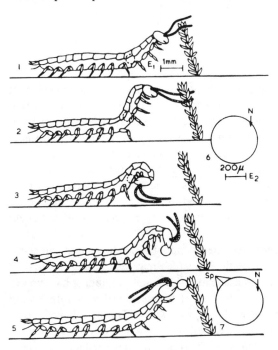

Fig. 45. Female symphylid having 'eaten' a spermatophore and deposited the sperm in a special cheeck pouch fertilizes each ovum by smearing sperm over it. *1* and *2* the female looks for a suitable plant stalk; *3* she removes the ovum from her sexual aperture in the third segment of the body, *4* and *5* she sticks the ovum to the plant; *6* and *7* ovum greatly magnified *Sp* sperms on the surface of the ovum; *N* nucleus. (Cloudsley-Thompson 1976 after L. Juberthie-Jupeau)

5.3.2 Uniramia

As in the Chelicerata, indirect spermatophore transfer is characteristic of the more primitive representatives of the terrestrial mandibulates. In Scolopendromorpha the male spins a web on which, after highly complicated pre-mating behaviour, he deposits a spermatophore to which the female is guided by threads of the web. Similar behaviour occurs in the Geophilomorpha and Lithobiomorpha. The males of Scutigeromorpha play an even more active, but primitive, role, transferring spermatophores with their mouths into the genital tracts of the females.

Male Pauropoda spin primitive webs in the absence of females, in one which they deposit two sperm drops. These webs probably attract the females and act as releasers for them to take up the sterm drops. Indirect free sperm transfer by droplet spermatophores has probably evolved from indirect spermatophore transfer, which is why it is mentioned here.

Symphyla are especially interesting because they have been considered to be the basic group from which the Hexapoda originated (Chap. 4.1). The males produce primitive stalked sperm drops which are taken up directly by the females. These sperm droplets are strikingly similar in structure to those of Diplura and Collembola. Female symphylans bite off the tops of these spermatophores when they find them, and store the semen in special gnathal sperm pockets. The eggs are fertilized as they are laid; the female pulls them from her genital atrium, one after the other and, with her mouth parts, deposits drops of semen in them (Figs. 44 and 45).

5.4 Indirect Sperm Transfer

5.4.1 Arachnida

Indirect transfer of sperm is found in some arachnid groups, and has been derived from indirect transfer of spermatophores via the substrate. Its origins can be seen in the transfer of spermatophores by the forelegs of male Cheliferidae, which insert them into the genital opening of the female. In Solifugae, no true spermatophore is formed, but a mucilagenous mass of sperms, or sperm ball, is transferred by the chelicerae of the male into the vagina of the female. The males of Ricinulei use their third pair of legs to transfer sperm, for which function the tarsal and metatarsal segments have been modified. Several different methods of sperm transfer appear to have been evolved independently in mites, including insemination by means both of the chelicerae and the third pair of legs.

The ancestral spiders probably deposited spermatophores on the ground and transferred them by means of claws on the pedipalps, which may also have been used to hold open the genital orifices of the females. From this simple beginning has evolved the complex palpal insemination that we find in male spiders today. Before mating takes place, the male spider charges his pedipalps with sperm. During this process of 'sperm induction' he spins a tiny triangular web with silk from his posterior spinnerets, on which a drop of seminal fluid is deposited. This is drawn into a duct within the pedipalps perhaps by resorption of a fluid previously secreted by surrounding glands. In a few spider families, both pedipalps are inserted into the sperm drop simultaneously; in Mygalomorphae and most Araneomorphae, however, the palps are inserted alternately. After courtship, one of five basic mating positions is taken up and the sperm introduced into the vagina of the female.

In the first of these, position I, the male stands with his body at an angle to that of the female; or with his dorsum opposed to her venter, but facing in the opposite direction. This position is found in Mygalomorphae and many Haplogynae. In position II, which is taken up in hunting spiders and some Agelenidae, the male mounts the back of the female, again facing in the opposite direction. In position III, which is characteristic of some Thomisidae, Araneidae and Theridiidae, the male crawls beneath the female and takes up a venter-to-venter position, but facing in the same direction as the female. In crab-spiders, position III is derived from position II, in web-building species from position I. Two other positions have been described, but are rare. In one of these the male and female are venter-to-venter and facing in oppositions. Sometimes the female lies on her side while the male stands over her (Fig. 46).

Loss of the spermatophore and rentention of the pattern of deposition of spermatic fluid in spiders are believed to be correlated with the evolution of web-spinning. A mass of sperm merely deposited on the ground would be likely to be absorbed between the soil particles before it could be taken up by the pedipalps. The problem is, however, overcome by spinning a web which does not absorb semen. This conclusion is supported by recent work which suggests that all araneomorph spiders, including ground-hunting families, are descended from web-spinning ancestors.

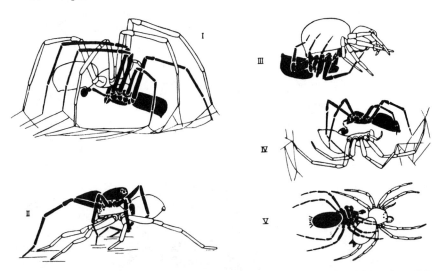

Fig. 46. Mating positions of spiders: males, black; females in outline *I Linyphia; II Pardosa; III Xysticus; IV Chiracanthium; V Antistea.* (J.L. Cloudsley-Thompson after B.J. Kaston from various authors)

5.4.2 Insecta

Indirct sperm transfer by Odonata has evolved, not from intromittent copulation, but from direct spermatophore transfer. The male dragonfly grasps the thorax of the female with his second and third pair of legs, while his first pair touches the bases of her antennae. He next flexes his abdomen forward and fits his claspers (two pairs on the tenth abdominal segment) into position on the female. Then he lets go with his legs and the two insects fly 'in tandem' (Fig. 47). The superior claspers of Anisoptera fit round the neck of the female, while the inferior claspers press down on the top of her head: in Zygoptera the claspers grip a dorsal lobe of the female's pronotum.

In addition to claspers on the tenth abdominal segment, the accessory genitalia of dragonflies include intromittent organs on the second and third abdominal segments. In Anisoptera there is a three-segmented penis with an orifice on its convex surface and lateral accessory lobes which hold the tip of the female's abdomen during intromission. The penis of the Zygoptera is simple, its only communication is with the body cavity and there is no distal aperture.

Sperms are transferred by the male, from his terminal gonoduct to the vesicle of the accessory genitalia, by bending his abdomen forward. Some species do this before grasping a female; others do so afterwards. The female then brings her own abdomen forward to make contact with the accessory genitalia of the male. Some species of dragonfly copulate in flight, completing the process in about 20 seconds. Others settle before copulating. In the latter cases, copulation may continue for an hour or more.

The Thysanura, the aedeagus or penis is used to deposit spermatophores on the ground. If, instead, the spermatophore were to be deposited on the male's own body, then the second and third abdominal segments would be those most easily reached.

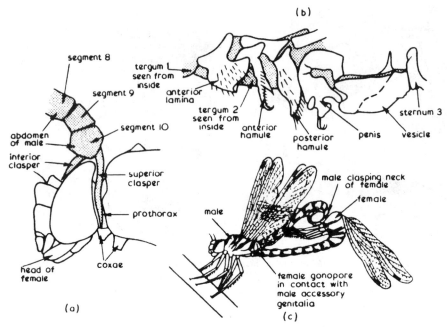

Fig. 47a–c. Mating in Odonata. a Position of the male claspers round the neck of the female *Aeshna* sp. b Male accessory genitalia of *Onychogomphus* sp.; terga of the left side removed. c Male and female *Aeshna* sp. in cop. (Cloudsley-Thompson 1976 after R.F. Chapman)

From this, one might postulate that special accessory genitalia to hold the spermato-phore could have been evolved and that these would develop into the condition found among modern Odonata. Although a 'copulation wheel' is formed in flight by Ephemeroptera, as well as by Odonata, the spermatophores of mayflies are trans-ferred while the abdominal terminalia are placed in apposition.

5.5 Direct Copulation with Free Sperm

Copulation with the exchange of semen must have evolved from copulation using spermatophores on numerous occasions. It is not uncommon in aquatic Crustacea and terrestrial mandibulates, but is less frequent among chelicerates other than Opiliones. Species with internal fertilization usually copulate, either by direct apposi-tion of their genital openings, or by means of various thoracic or abdominal appendages of the male. Terrestrial Decapoda must have evolved the venter-to-venter position before coming on land.

Harvest spiders enjoy copulation with the aid of an intromittent organ of inor-dinate length. Indeed, the penis may measure up to 3.5 mm, thus actually exceeding the linear dimensions of the body. This penis consists of a non-segmented projection of the area round the gonopore. It terminates in a recurved stylet, at the end of which

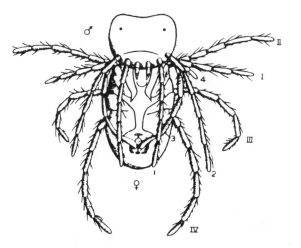

Fig. 48. Direct sperm transfer in a mite, *Piona*. *Arabic numerals* male; *roman numerals* female. (Cloudsley-Thompson 1976 after F. Schaller)

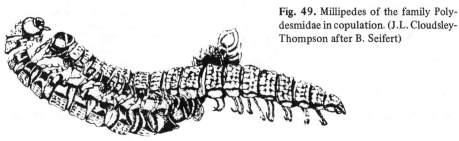

Fig. 49. Millipedes of the family Polydesmidae in copulation. (J.L. Cloudsley-Thompson after B. Seifert)

opens the ejaculatory duct. There is practically no courtship but, in *Mitopus morio,* the male has been observed to run towards a moving female and take up a position with his body just above and with legs straddling hers. The two may run together in this position for several centimetres before stopping. The male then moves forward slightly, so that his body is now in front of that of the female, and turns round to face her. His long genitalia are now thrust forward and mating takes place. In this case, the function of courtship is clearly to prevent escape of the female (Sect. 5.2.2).

Spermatophore transfer in mites (Acari) may take place by means of stalked spermatophores (Oribatei), the chelicerae (Mesostigmata), or the third pair of legs acting as gonopods (Hydrachnellae). Copulation by means of a penis occurs in Acaridei (Fig. 48). During mating, the males often hold their females clasped in a tight embrace. In many bird parasites (e.g. *Analges*) the males' legs are modified for this purpose. Suction cups or cement are also sometimes used to secure the female mite during mating. These cups are usually situated on the posterior part of his body so that the male faces away from his mate while copulating.

In most millipedes, the male is endowed with legs modified to form gonopods by which sperm is transferred directly to the vulva of the female (Fig. 49). These gonopods form a useful diagnostic character and enable males to be recognized at a glance. In the African pill-millipede (*Sphaerotherium*), however, there are no gonopods: the male uses normal walking legs for the transfer of sperm. Clutching the anterior vulvae

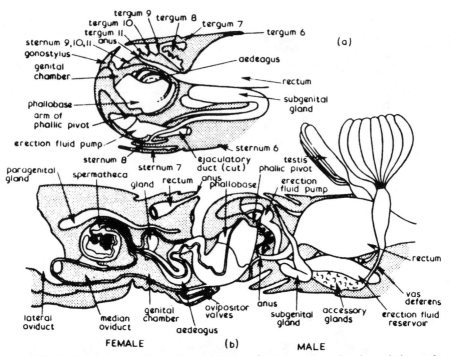

Fig. 50a,b. Copulation in *Oncopeltus fasciatus*. **a** Sagittal section of male genital capsule with aedeagus retracted. **b** Sagittal section of the posterior ends of copulating *O. fasciatus* showing inversion of the male genital capsule and insertion of the aedeagus into the spermatheca. (J.L. Cloudsley-Thompson after R.F. Chapman)

of the female with his pincer-like terminal legs, the male secretes a large sperm drop from his anterior genital pores. This is then passed posteriorly, from the tarsi of one pair of legs to those of the next, until it reaches the female genital opening. That is probably the most primitive case of sperm transfer by unspecialized appendages known in terrestrial animals.

In other Glomeridae, the males clutch the vulvae of the females with their telopods and then roll up to clean their terminal legs. Next, they grasp a particle of substrate (usually a pellet of frass) with their forelegs, roll it on the ground and gnaw it until it is spherical. The particle is then applied to the genital pores and a drop of sperm thereby transported indirectly to the telopods which introduce it into the vulva of the female.

Sperm transfer is indirect in Pselaphognatha. It is considered here, however, as it is the final link in an evolutionary sequence from direct copulation to indirect sperm transfer. The male *Polyxenus lagurus* spins a double zig-zag thread, above a small depression, with his 'penis appendage', and places two drops of sperm on it. He then moves away, spinning a couple of parallel threads as he retreats. The threads contain a pheromone which attracts sexually mature females so that they follow them to the sperm droplets which are taken up with the vulvae. The signal threads also stimulate other males, which eat the old sperm drops and replace them with new ones, thus not

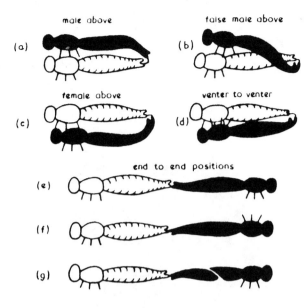

Fig. 51a–g. Different positions assumed by male and female insects during copulation: male black; female white. **a** Male above (e.g. some Diptera). **b** False male above (e.g. Acrididae). **c** Female above (e.g. some Orthoptera). **d** Venter-to-venter (e.g. Diptera Culicidae – hypopygium inverted). **e** End-to-end male abdomen not twisted (e.g. some Hypenoptera). **f** End-to-end male inverted (e.g. some Tettigonioidea). **g** End-to-end male abdomen twisted (e.g. some Heteroptera. (J.L. Cloudsley-Thompson after R.F. Chapman from various authors)

only economizing in protein and nucleic acid, but also providing constantly fresh sperm for the females.

Pterygote insects copulate using either free sperm or semen contained in spermatophores. For example, many Hemiptera produce spermatophores. In *Rhodnius prolixus* these consist of a gelatinous mass of mucoprotein originating from the accessory glands of the male and solidifying in the more acid environment of the intromittent organ. Retained in the bursa copulatrix of the female, the spermatophore functions as a sperm plug, holding the semen until it has been sucked into the spermathecae by contractions of the female ducts. Another trend in the Heteroptera is exemplified by *Oncopeltus fasciatus.* In this species, semen is deposited directly in the spermatheca by the long, flagellar, intromittent organ of the male (Fig. 50). The spermatophore is absent in *O. fasciatus,* but accessory glands are present as apparently functionless mesadenes whose secretion has many histochemical properties in common with the secretion of the glands in *R. prolixus.* This evidence supports the view that the spermatophore is a primitive character and that insects delivering free semen have evolved from forms that produced spermatophores.

All pterygote insects copulate, but other labiates, apart from Diplopoda, transfer sperm droplets or spermatophores indirectly. Copulation in insects may have evolved only once; but, with the possible exception of Acari, more variations in the manner of copulation occur among the Pterygota than among all other Arthropoda. Of the various insect copulatory positions known (Fig. 51), female-above tends to be the most primitive. Because intromittent organs are restricted to species with male-above or end-to-end positions, any other explanation would need to account for the loss of intromittent organs, an event possibly unknown among any animals. Venter-to-venter copulation with both partners facing in the same direction probably evolved as a primitive mating position only in aquatic Crustacea and terrestrial Diplopoda. Among the Pterygote groups shown in Table 1, only the Tettigonioidea and Blattoidea include

Table 1. Copulatory positions of Pterygota

Pterygote insect groups	Approx. age of oldest fossils (Millions of years)	Variation in copulatory positions				End-to-end			
		♀-above (♀ Venter to ♂ dorsum)	♂-side (♂ Reaching under ♀)	False ♂-above (♂ Reaching under ♀)	♂-above (♂ Venter to ♀ dorsum)	♂ Abdomen slightly or not twisted	♂ Abdomen strongly twisted	♂ Inverted (Lying on back)	Venter to venter (Facing same way)
Siphonaptera	50	×							
Diptera	160	×	×	×	×	×	×	×	×
Lepidoptera	50		×			×	×		
Trichoptera	150	×	×			×			
Mecoptera	275		×				×		
Hymenoptera	160		×	×	×	×	×		
Coleoptera	240		×	×	×	×	×		
Neuroptera	275	×	×	×			×		
Thysanoptera	240						×		
Anoplura	1	×							
Mallophaga	.	×							
Psocoptera	275	×					×		
Homoptera	220		×	×	×	×	×	×	
Hemiptera	275		×	×	×	×	×	×	
Plecoptera	240		×	×					
Embioptera	50		×	×					
Phasmodea	60		×	×					
Acridoidea	60		×	×					
Tettigonioidea	300	×				×	×	×	
Isoptera	50						×		
Dermaptera	160			×			×		
Mantodea	50						×		
Blattodea	320	×	×			×			
Odonata	275				×				
Ephemeroptera	275	×							

Probable transitional Relationships

species lacking intromittent organs and, in all case, elaboration of a penis appears to be related either to a male-above or to some kind of linear position in which the male is the dominant partner. A notable exception is found in the Mecoptera (*Boreas*), but here the male holds the female on to his back with tong-like wings and there is an elaborate intromittent organ but no spermatophore.

As far as is known, the orthopteroids, Ephemeroptera, and Lepidoptera probably possess spermatophores in all species; the Mecoptera, Diptera, and Siphonaptera lack them in all species. Other groups appear to have some species with spermatophores and others without.

5.6 Haemocoelic Insemination

Fertilization is internal among Onychophora but *Peripatus* deposits spermatophores anywhere on the body of the female, and hypodermic impregnation takes place. The same occurs in *Opisthopatus*. Among rotifers and some helminths, the spermatozoa are injected by the penis of the male directly into the body of the female. The only arthropods to do the same thing are bugs of the super-family Cimicoidea. Stages in its evolution are to be found in Nabidae, Cimicidae, Plokiophilidae and Polyctenidae.

In some Nabidae, spermatozoa are deposited in the genital chamber of the female. Some of them penetrate the glandular walls of the oviducts and, passing through the peritoneal coat and between the fibres of the surrounding muscles, escape into the blood, where they are quickly phagocytosed. The next step occurs in certain Cimicidae. Spermatozoa are injected directly into the haemocoel through the walls of the vagina or genital chamber. Many are destroyed, but some succeed in reaching the oviducts and ovarioles. The absorption of spermatozoa may be taken over by spermaleges. These are areas of the female abdomen specialized for receiving the copulatory organs of the male. In *Prostemma* (Nabidae), for example, spermatozoa are injected through the vaginal walls, and the mesodermal part of the spermalege (mesospermalege) is very developed. It is derived from the fat body but, more usually, spermaleges originate from blood cells.

In *Primicimex cavernius*, a bug that inhabits bat caves in Texas, spermatozoa are injected into the haemocoel through the outer integument of the female. At each copulation she is pierced in a different place and spermaleges are not found. The left clasper of the male acts as a sheath for the penis, which perforates the cuticle of the female. The spermatozoa are distributed to all parts of the body of the female, and stored in pouches at the base of the oviducts. Ectospermaleges, specialized to receive the copulatory organs of the male, have evolved independently in different genera of Cimicidae and Polyctenidae. The introduced spermatozoa are subsequently absorbed by mesospermaleges. The final development is that the mesospermalege becomes not only responsible for the phagocytosis of much of the semen, but conducts part of it to the ovaries. In *Pagasa* (Nabidae) the mesospermaleges form only rudimentary conducting tissue; but an extremely specialized stage of haemocoelic insemination has been evolved independently in the Cimicinae and Anthocorinae.

The adaptive function of haemocoelic insemination is not clear since it always results in digestion of the seminal fluid and of some of the spermatozoa injected.

H.E. Hinton (1964) has argued that it has selective value in providing the females with protein which helps them to survive long periods of starvation. In *Afrocimex* spp., however, males have ectospermaleges and inseminate one another. Females possess both ectospermaleges and mesospermeleges. The only benefit to a male bug that accrues from insemination by another would appear to be in the protein that it obtains thereby. The populations, however, may benefit from the fact that this habit will increase variations between individuals in the length of time that they can withstand starvation. It is not known whether this should be interpreted as implying group selection, or whether the populations concerned are sufficiently closely related to implicate kin selection. Until population sizes have been estimated, it is probably wise to suspend judgement on this point.

5.7 Conclusion

Most sexual differentiations in arthropods are phylogenetically recent, as are the remarkable alterations that occur in the positions of their genital apertures. Progoneate and opisthogoneate forms are found not only in myriapods but also in chelicerates and some of the Hexapoda. Although sperm transfer by means of spermatophores was used in all taxa during the transition from aquatic to terrestrial life, it is by no means clear whether the spermatophores of crustaceans, arachnids, myriapods and insects are homologous or not. Adaptation to life in soil has led to many parallel developments resulting from convergence. Thus, although homologies have been postulated between the stalked spermatophores of scorpions, pseudoscorpions, Thelyphonida and Phrynichida, the striking similarities between the spermatophores of Collembola and of oribatid mites cannot be regarded as anything but analogies. At family and generic level, however, modes of sperm transfer can be used with confidence to clarify problems of relationship.

The morphological and functional specialisations of Solifugae, Ricinulei and Araneae have clearly evolved in parallel, and independently of one another. So, too, must the evolution of a true penis in male Opiliones and certain families of mites. Indeed, different families of Acari exhibit almost all the different modes of sperm transfer that occur throughout the other orders of Arachnida. The systematic occurrence of these different modes, therefore, does not lead to any conclusions concerning their evolutionary significance. On the contrary, analogous mechanisms of sperm transfer appear to have been evolved independently in different taxa in response to ecological and environmental factors.

Further Reading

Alexander AJ, Ewer DW (1957) On the origin of mating behavior in spiders. Am Nat 91:311–317
Alexander RD (1964) The evolution of mating behaviour in arthropods. Symp R Entomol Soc Lond 2:78–94
Bastock M (1967) Courtship. A zoological study. Heinemann, London
Bristowe WS (1958) The world of spiders. Collins, London
Chapman RF (1969) The insects, structure and function. English Universities Press, London

Cloudsley-Thompson JL (1976) Evolutionary trends in the mating of Arthropoda. Meadowfield, Shildon Co Durham (Patterns of Progress Zoology, Vol 5)

Davey KG (1960) The evolution of spermatophores in insects. Proc R Entomol Soc Lond (A) 35:107–113

Hinton HE (1964) Sperm transfer in insects and the evolution of haemocoelic insemination. Symp R Entomol Soc Lond 2:95–107

Lewis JGE (1981) The biology of centipedes. Cambridge University Press, Cambridge London

Schaller F (1971) Indirect sperm transfer by soil arthropods. Ann Rev Entomol 16:407–446

Schaller F (1979) Significance of sperm transfer and formation of spermatophores in arthropod phylogeny. In: Gupta AP (ed) Arthropod phylogeny. Van Nostrand Reinhold, New York, pp 587–608

Thornhill R, Alcock J (1983) The evolution of insect mating systems. Harvard University Press, Cambridge Mass London

Weygoldt P (1969) The biology of pseudoscorpions. Harvard University Press, Cambridge Mass

6 Adaptations to Extreme Environments

The land masses of the world provide far more extreme and variable habitats than do the oceans. It is to such environmental extremes that terrestrial arthropods must become adapted if they are to survive. This they can do either behaviorally, avoiding the most severe conditions by retreating into crevices and holes, or by physiological means, or by both. Physiological adaptations may occur in some or all stages of the life cycle. The most extreme environmental conditions are sometimes resisted in a state of diapause (Sect. 6.6.1) or even in one of cryptobiosis (Sect. 6.6.2). Examples of physiological adaptation to extreme environments are afforded by the fly *Psilopa petrolei* which inhabits puddles of crude petroleum, feeding on the dead insects found there, and by the beetle *Niptus hololeucus,* which can live on cayenne pepper and thrive on sal ammoniac. This species has also been found inhabiting the corks of entomologists' cyanide killing bottles. Tenebrionidae, including the mealworm beetle *Tenebrio molitor,* can live on dry food without drinking, although excessive dryness in a mealworm culture leads to cannibalism; while the larvae of carpet beetles, clothes' moths and other arthropod pests of stored products can also do without free water completely.

6.1 Desert Adaptations

The desert biome presents, to the ultimate degree, all the problems of terrestrial life. Its extreme physical and climatic conditions have engendered in arthropods a number of inter-related morphological, behavioral, and physiological adaptations, many of which are parelleled in various other unrelated taxa of animals and plants (Figs. 52, 53). The integuments of desert insects and the cuticles of desert plants both have extremely impervious wax layers, while insect spiracles and plant stomata are especially efficient among desert species. There is a positive correlation between the high critical temperatures of arthropod cuticles and the high ambient temperatures that the animals experience in their natural habitats. Xeromorphic desert plants are characterized by having the stomata buried in the epicuticle below the general level of the plant surface, or superficially depressed due to the extremely thick surrounding cuticle, while the spiracles of desert arthropods are likewise often sunken or hidden below the surface of the body. Furthermore, lack of vegetation and exposure to predation have intensified the effects of natural selection, and superimposed yet another stratum of adaptive characters.

Insect integument   Plant epicuticle

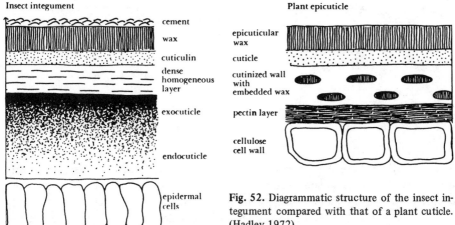

Fig. 52. Diagrammatic structure of the insect integument compared with that of a plant cuticle. (Hadley 1972)

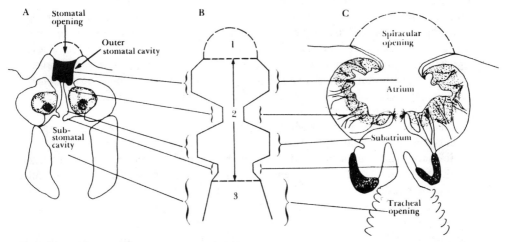

Fig. 53A–C. Comparison of diffusion resistance within and outside a plant stoma and insect spiracle. **A** Cross-section of stoma of *Yucca;* **B** schematic representation of relative resistance to diffusion through an aperture of varying diameters *1* external resistance (boundary layer); *2* stomatal (spiracular) resistance; *3* substomatal (tracheal) resistance. **C** Horizontal section through left spiracle of a beetle. (Hadley 1972)

6.1.1 Morphological Adaptations

The morphological adaptations of Arthropoda to the desert environment fall into two broad classes: adaptations that reduce water loss by transpiration, and adaptations for moving across, or burrowing into, sand. The lung-books and spiracular closing mechanisms are especially well developed, as we have seen.

Adaptations of scorpionid scorpions for burrowing include massive pedipalps, reduced metasoma, and small pectines with comparatively few teeth. Some of the

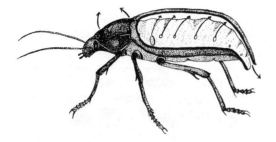

Fig. 54. Diagram of a desert tenebrionid beetle showing opening of spiracles into the subelytral cavity and avenues of water loss. (After Ahearn 1970)

Buthidae, which have slender claws, are surprisingly good burrowers also, apparently using their legs to a greater extent than do the Scorpionidae. Solifugae use their chelicerae to bite at the substrate, and rake loosened particles back under the body with the second or, less often, the third pair of legs. The chelicerae, sometimes assisted by the pedipalps and legs, are used like ploughs to push excavated materials from the burrow.

Many desert spiders likewise inhabit burrows and, whether Mygalomorphae or Araneomorphae, are usually large, with relatively heavy bodies and thick-set legs. Some have brushes of hair on the undersides of their limbs, which facilitate movement on sand. Species of *Aganippe* in Australia have a dense scopula on the lower sides of their front tarsi and metatarsi, while the dune-dwelling *A. simpsoni* has rigid brushes of vertical bristles on the last joints of the palps and front legs: these are presumably used to push encroaching sand away from the burrow entrance. Paradoxically, sheet flooding is a great hazard in many arid regions, and Australian mygalomorphs have evolved a diversity of antiflooding structures which either seal the burrow against inundation, or else deflect water away from the entrance.

In xerophilous buprestid beetles, the spiracular openings are covered by a basket-work of outgrowths. These are believed to impede the diffusion of water molecules to a greater extent than those of carbon dioxide or oxygen. The abdominal spiracles of desert Tenebrionidae open into the subelytral cavity (Fig. 54) rather than directly to the atmosphere. This significantly reduces respiratory water loss, which increases considerably when the elytra are removed. It may also, to some extent, assist in thermoregulation.

6.1.2 Physiological Adaptations

Physiological specializations common to most desert organisms include tolerance of low and high temperatures, facultative hyperthermia, relatively low cuticular and respiratory transpiration, efficient nitrogenous excretion, atmospheric water uptake, conservation of metabolic water, and resistance to desiccation. Although no single species exhibits all these features, many of them are to be found among different orders of Arthropoda. Scorpions and Solifugae are again the most highly adapted physiologically: water obtained from the food is conserved, however, and the Acari is the only arachnid order in which active uptake of moisture from unsaturated air has so far been demonstrated, but it occurs in the Mojave desert cockroach *Arenivaga* and in Namib desert Thysanura. Whether water is absorbed through the cuticle, the rectum, or both, has not been fully established. Nor has the physiological mechanism involved, although various explanations of the active transport of water vapour have been made.

Water vapour can be absorbed from unsaturated air by various mites and ticks, Thysanura, flea prepupae, larval and female American desert cockroaches (*Arenivaga investigata*) and certain other wingless insects — only the male *Arenivaga* has wings. The adaptive significance of this ability is obvious: its apparent lack in desert Tenebrionidae may reflect their extremely low rates of transpiration. (On the other hand, there could well be a region of the abdomen that absorbs water vapour from the sub-elytral cavity). The mechanism involved is still not known, and the site of absorption has not been determined with certainty. There is evidence that it may be the rectum in some insects, but this is not so in Acari.

Loss of water accompanying the elimination of nitrogenous waste is minimized in arachnids by the excretion of insoluble guanine as the end-product of protein catabolism and, in insects, of insoluble uric acid (Chap. 2.7.2). Whereas the osmolarity of the haemolymph is regulated in desert Tenebrionidae, scorpions tolerate increasing osmotic pressures when subjected to desiccation. Heat death is associated with a decrease in the pH of the blood and may be related to the accumulation of acid metabolites at high temperatures faster than they can be eliminated.

Most desert insects and arachnids are unable to tolerate any greater loss of water than are their relatives from more humid places, although they are often more resistant to water loss by evaporation. For example, the American desert scorpion *Hadrurus arizonensis* loses only 0.028 per cent of its body weight per hour in dry air at 30°C. The Sudanese camel-spider *Galeodes granti* loses 0.147 per cent at 33°C, but is unusual in that it does not die until it has lost as much as two-thirds of its total weight.

Metabolic water, produced by the oxidation of food reserves within the body, contributes to the water economy of all animals and may be adequate for some Tenebrionidae. In the desert locust *Schistocerca gregaria* and most other desert arthropods, however, it is insufficient to maintain water balance and preformed water is obtained from the food or, in cool coastal deserts such as the Atacama and Namib, from condensed fog.

During desiccation, the osmotic pressure of the haemolymph is regulated by the excretory system. During dehydration in the absence of food, however, additional mechanisms, perhaps involving exchanges of water between the haemolymph and the tissues, may be necessary to maintain the output of solutes required for osmotic control. In the presence of food, the main problem lies in determining the way in which the preformed water is absorbed If large amounts of salts are also ingested, the excretory capacity of the malpighian tubules may be exceeded and the osmotic pressure of the haemolymph rises. The rate of salt uptake by the midgut may also be regulated physiologically.

6.1.3 Behaviour and Rhythmic Activity

Most desert arthropods escape the high temperatures and extreme saturation of the desert surface by burrowing or seeking shelter during the day, and restricting their activities to the hours of darkness. This behaviour is regulated by their circadian 'biological clocks'. These are endogenous, self-sustained oscillations which persist under constant conditions, with a periodicity of approximately 24 h, and can be entrained by environmental synchronizers, of which light is by far the most important. Desert Arthropoda tend to be more strictly nocturnal in habit than related species from more humid biomes. This may be correlated with the avoidance of climatic extremes, predatory enemies, or both. Desert scorpions and Solifugae are unusually resistant to high temperature, and have very low transpiration rates. For this reason, it seems probable that the avoidance of predators may be more significant than thermal physiological requirements in their night-active behaviour, as well as in that of desert Thelyphonida and the larger mygalomorph spiders. Despite their venomous stings, several species of reptiles, birds and mammals are known to feed on scorpions. Furthermore, scorpions come out into the open far less frequently in bright moonlight than they do on dark nights.

Desert arthropods are not only more strictly nocturnal than temperate and tropical forest species, but they are generally much more active. Two alternative predatory strategies are available to large, tropical Arthropoda. They may remain relatively inactive within their sheltered burrows and retreats, or else they can venture forth in search of prey. Forest and woodland centipedes, scorpions and mygalomorph spiders tend to adopt the first of these: they often spend long periods of time comparatively motionless. What movement they do show is far less rhythmic than that of desert species, which are strictly nocturnal and much more active. A correlation is evident between the type of habitat, the amount of timing of locomotory activity, and the rate of transpiration. Day-active desert beetles, such as some Tenebrionidae and Scarabaeidae are, in general, distasteful and their black colours have an aposematic function (Chap. 8.3).

The arthropod fauna of the desert can be subdivided roughly into two ecological groups: (a) long-lived species including scorpions, Thelyphonida, large tube-dwelling Mygalomorphae, a few of the larger Lycosidae which also inhabit burrows and some Sparassidae; tenebrionid and scarabaeid beetles etc. (b) Solifugae and small nomadic hunting spiders of the families Gnaphosidae, Salticidae and Thomisidae, Thysanura, Orthoptera, Lepidoptera, Diptera, Hymenoptera etc., whose life span does not exceed one year. The latter may represent r-selection forms, in an r-K continuum. The more important correlates of r-selection are associated with an unstable environment, unpredictable resources, catastrophic mortality, highly variable population size, rapid development, a short life cycle, small size and high productivity. K-selection, on the other hand, is mostly correlated with the opposite conditions and is exemplified by scorpions and the other arachnids included in group (a) above.

Large mygalomorph spiders, numerous in the Australian and North American deserts, are virtually absent from the Great Palaearctic desert where their ecological niche is apparently occupied by Solifugae. These are far more numerous and diverse in the Old World than they are in the New. They do not occur at all in Australia, where rainfall is extremely erratic, probably because they are r-selected and live for only one year. The annual movement northward of the inter-tropical front imposes some regularity on the summer rains of the southern fringes of the desert regions of Africa, Asia and North America. This may enable Solifugae to compete there on favourable terms with the K-selected Mygalomorphae.

6.2 Forest Adaptations

The arthropods of the forest biome can be divided into two ecological groups: the cryptozoic inhabitants of the soil and leaf litter, and the terrestrial and arboreal forms that live in more exposed localities.

6.2.1 Cryptozoa

Cryptozoic animals lead hidden lives, as their name indicates. They are extremely numerous, comprise a distinct community of animals, and the fauna of the forest floor in tropical and temperate regions is rich in species. Many important orders and families of cryptozoic animals are to be found only in forest humus and leaf litter. Indeed, scorpions and Solifugae provide rare examples of cryptozoic arthropods that are better represented in savanna and desert than they are in forest.

The cryptozoic arthropod fauna of forest humus includes Onychophora, amphipods, woodlice, Paurpoda, Symphyla, millipedes, centipedes, scorpions, false-scorpions, Ricinulei, harvest-spiders, spiders, mites, Collembola, Protura and, of course, insects (Fig. 55). Although cryptozoic arthropods tend, in general, to be small, some tropical forms are considerably larger than their relatives in the temperate regions of the world.

The humus of the forest floor not only provides shelter for its numerous inhabitants but is itself sheltered by the trees and shrubs growing above. The cryptozoic fauna of forests is therefore doubly protected from the regours of the environment. Thermal and humidity fluctuations are almost eliminated and light is excluded. The drab uniformity of structure and colour of such animals stands in marked contrast to the enormous diversity of shape and hue found amongst the Diptera, Coleoptera, Lepidoptera and other typical arthropods of the open air. Whereas the cryptozoic fauna of the forest is limited by the physical conditions of the environment, and cannot exist beyond its confines, the creatures of the open are very largely released from such restrictions, and are adapted to a variety of environmental conditions. Nevertheless, it may be the case that less humid terrestrial environments of the earth were originally colonized by forms that originated from the damp soil and humus of tropical forests (see below).

Cryptozoic arthropods are characterized by their lack of specialization. They are usually extremely dependent upon a moist environment, and die rapidly from desiccation when removed from it. They tend to belong to classes whose members do not normally possess discrete epicuticular wax layers such as woodlice, centipedes and millipedes, so that water loss by evaporation tends to vary according to the satura-

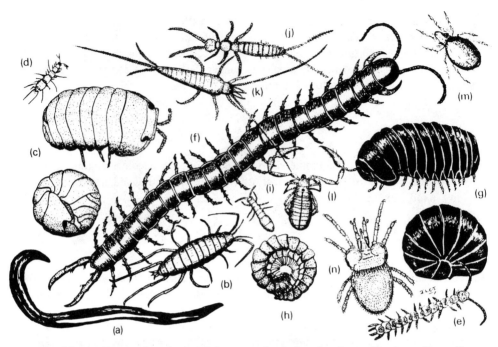

Fig. 55a–n. Cryptozoic animals of forests: **a** Land-planarian; **b** amphipod; **c** pill-woodlouse; **d** pauropod; **e** symphylan; **f** centipede; **g** pill-millipede; **h** colobognath millipede; **i** collembolan; **j** dipluran; **k** thysanuran; **l** false-scorpion; **m** oribatid mite; **n** trombiculid mite. (Not to scale). (Cloudsley-Thompson 1969)

tion deficiency of the atmosphere. Even taxa that do possess an impervious layer of wax in their integuments tend to have higher transpiration rates than do related forms from less humid environments (Chap. 2.2). Such forms are primitively nocturnal, too, and emerge from their hiding places at dusk when the temperature drops and the relative humidity of the air increases, but they are not so markedly nocturnal as desert species, as we have seen (Chap. 6.1.4).

Other distinctive characters of cryptozoa, as compared with other groups of animals, include: (a) small size and lack of pigmentation; (b) lack of efficient respiratory mechanisms which are either absent or, when present, lack devices that control water loss by evaporation; (c) poor development of visual sense organs, associated with well-developed tactile and taste sensillae; (d) secondary sexual characters distinguishing males from females are much reduced, and there is usually little in the way of outward appearances to separate the sexes; (e) development from eggs that do not possess an impervious shell and are therefore laid in clusters coated with mucus. In many orders, the delicate, thin-skinned young are retained for a while after hatching in a brood-chamber or its equivalent. During this period, the helpless young remain with the mother, and are covered with the secretions that she produces until they are able to lead an independent existence; (f) moulting often continues throughout life, even after maturity has been reached and growth ceases. Striking examples are afforded by the Onychophora and Diplopoda. Many cryptozoic animals eat their discarded integuments, a type of behaviour not practised by other land-adapted arthropods. This habit may be related to the shortage of calcium in tropical forest soils; (g) most cryptozoa are inactive, slow-moving animals.

The family history of the cryptozoa has greater continuity and reaches back longer in time than does that of the present inhabitants of open lands. Many cryptozoic orders of insects were already established in the Palaeozoic period, and only a minority have failed to survive until today, while primitive and archaic forms such as Onychophora, Pauropoda, Symphyla and Palpigradi are well represented. All this suggests a community of animals which, in former times, was the product of an environment more universally simple and uniform than it is today. Evolutionary progress has died away in the absence of environmental changes, leaving the cryptozoa imprisoned in a habitat from which no further advance can be made and from which escape is forever impossible.

6.2.2 Flying Insects

Flying insects may not exhibit any marked adaptations for arboreal life, but the richness in species of tropical and temperate forests reflects the luxuriance of the vegetation that supports them. The fauna of trees is largely divided between the foliage and the trunks. Forms that depend directly upon foliage may be surface feeders, leaf-rollers, leafminers, or gall formers. Those of the tree trunks may bore into the cambium, cambium wood, or heartwood, or else they may dwell beneath the bark. A single oak tree can harbour 50,000 caterpillars; a single leaf, more than 5,000 aphids, but the greatest numbers, variation, and ecological complexity are found in the tropical rain forest.

For animals as small as arthropods the concept of arboreal adaptations is not applicable. The microenvironments associated with trees are far too heterogeneous to have engendered any specific adaptations. Insects may be leaf-eaters, pollinators, wood borers and so on — adapted to different parts of trees, but not to trees themselves.

6.3 Arctic and Alpine Adaptations

Arctic and alpine arthropods are subjected to many selective influences that are common to both types of environment. There are certain differences, however. At high altitudes, atmospheric pressure is reduced. Photoperiod is related to latitude rather than to altitude, and at high elevations the atmosphere is thin. Consequently, insolation is powerful on mountain tops, in contrast to arctic snowlands, where the rays of the sun are strongly filtered as they pass obliquely through the atmosphere. The biologically significant features of arctic climates relate to the seasonal and diurnal patterns of insolation. Towards the poles, insolation becomes restricted to the short summer and the contrasts between winter and summer are greater than they are on mountain tops at lower latitudes.

Arthropods of the arctic polar desert include flying insects and soil-dwelling Collembola and Acari. During the brief snow-free season, moist soils teem with invertebrates which comprise a whole ecosystem based on detritus-feeders, but with one or two levels of predators at several points in the chain. Dipteran larvae are abundant in every pool and in wet soil. They constitute a considerable share of the animal biomass accumulated during summer. The pollination of flowers is mostly carried out by insects in the tundra, but more by wind at higher latitudes. Diptera are more important pollinaters than bees.

Considerable research has been undertaken on the physiology of survival of arctic insects at low temperatures. In some cases, insects are able to escape the lethal effects of freezing by supercooling. The body temperature can drop as low as $-20°C$ before crystallization of the haemolymph takes place. This may be effected by the synthesis of glycerol and other cryoprotective compounds before hibernation. Other factors, however, and especially the elimination of nucleating agents from the alimentary canal, are of even greater significance. Water that is completely pure often supercools to an extremely low temperature before freezing takes place. If, however, there are any dust or other particles in it, these nucleating agents form the bases on which crystals of ice begin to form.

The alpine fauna, in many ways, is similar to that of the arctic. Many arthropods are dark-coloured with pronounced pigmentation. In addition to black and brown, other colours commonly encountered are red, orange and deeper tones of yellow. With the exception of cryptozoic forms, pale or colourless species are absent above the snow line. The function of this is probably thermal, the darker colours absorbing more radiation from the sun.

Reduction of wings, or a totally wingless condition, are characteristic of high altitude insects throughout the world. In the mountains of East Africa (Fig. 56) and Ethiopia, for example, many insects are flightless and tend to be restricted to particular massifs. Carabidae and Staphylinidae predominate. Wing atrophy can be associated with very diverse conditions including sex dimorphism, sedentary modes of life, myrmecophilous or termitophilous habits, parasitism, and so on. The explanation of aptery has been based, by analogy with the similar loss of flight in beetles of oceanic islands, on the winds experienced at high altitudes. It might, however, equally be sought in the cryptozoic habit of the animals when they shelter from other violent conditions on the surface of the ground − from excessive radiation, great saturation deficiency, and wide and rapid fluctuations of temperature.

Insects that live at high altitudes tend also to be smaller than those at lower levels. Not only is there a progressive reduction in mean body size with increase in elevation in most orders of high-altitude arthropods, but orders which, like Diptera and Collembola, include very small forms, gain increasing dominance at higher altitudes and replace almost all others in the nival zone above the permanent snow line. The same applies to linyphiid spiders: in the tropics these are numerous only at high altitudes.

There is a progressive general flattening and increase in the width of the body among Tenebrionidae, which is associated with an increase in the convexity of the elytra. As in desert species, this reduces water loss by transpiration through the spiracles (Fig. 54), while the reservoir of air within may also serve to insulate the body tissues from excessive heating through insolation in the rarified atmosphere. Furthermore, as

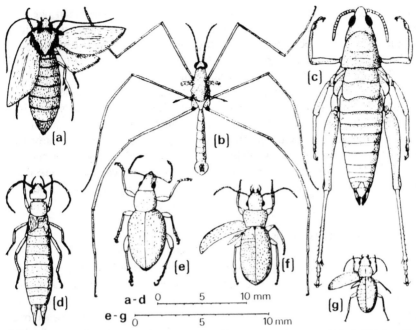

Fig. 56a–g. Flightless insects found on Kilimanjaro above 4250 m. a *Altia;* b *Tipula;* c *Parasphena;* d *Forficula;* e *Parasystiella;* f *Ploxamotrechus;* g *Peryphus.* (Cloudsley-Thompson 1969. After G. Salt)

elevation increases, there is an increasing predominance of species with dense bristles, setae, scales and waxy coverings, especially among species that frequent open snow and rock surfaces during the hours of bright sunlight.

Most arctic and alpine arthropods are characterized by pronounced stenothermy and are able to maintain activity when their bodies are really cold. They have a relatively low optimal temperature, usually around freezing point, and all their metabolic activities are adjusted to low temperatures. This varies, however, not only from species to species, but within different stages of the life-cycle of the same species. An insect that hatches from an over-wintering egg, or re-awakens from hibernation develops at increasingly higher temperatures with the advent of summer. There is, correspondingly, a gradual fall in cold stenothermy as the summer advances and the life-cycle progresses from the dormant to the active stages. In addition, the Arthropoda of nival regions become concentrated near the edge of the snow line, the ice margin or glacier snout, and the edges of glacial lakes and ponds. Only in moist situations can they expose themselves and absorb warmth from the direct sunshine without risk of desiccation.

Active life is severely restricted to the summer months, and seasonal rhythms are extremely well marked except on equatorial mountains. Here there is no pronounced seasonal rhythm but all activity ceases at dusk and is resumed again at sunrise. Prolonged hibernation under snow cover and a short period of rapid development are characteristic of high-altitude insects of temperate and arctic regions. Most species have short live-cycles which are completed within a year, so that hibernation always

takes place at the same stage of development. A few require longer than one year
to complete a single generation at high altitudes: in general, however, the environment favours univoltine species which pass through one generation per year. Insects
with facultative diapause may complete two generations per year, but those with
obligatory diapause (Sect. 6.6.1) are strictly univoltine.

It is curious that the arthropods which inhabit the most inhospitable places on
earth should be members of the oldest and most primitive orders. Perhaps there is an
analogy between the highest altitudes and latitudes and the sterile land of the Silurian
period when terrestrial arthropods evolved from aquatic ancestors. The early land
animals may have fed on wind-blown debris that accumulated among barren rocks of
the world beyond the fringes of shore plants, just as high-altitude Acari, Collembola,
Thysanura and so on subsist on pollen and plant debris blown up from lower levels.

6.4 Littoral and Aquatic Adaptations

Although marine Xiphosura, and the Pyconogonida are primitively aquatic, certain
other classes of arthropods, namely Pauropoda, Symphyla, Chilopoda, Diplopoda,
Collembola, Insecta and Arachnida contain species which have become adapted to life
on the sea shore. Intertidal arachnids include Palpigradi, Scorpiones, Pseudoscorpiones,
Araneae, and Acari. The principal marine insects are Thysanura, Hemiptera, Trichoptera, Coleoptera and Diptera. Even on the open ocean, thousands of kilometres from
land, can be found sea-skaters (Hemiptera: Hermatobatidae). Five species of the genus
Halobates spend their entire life on the sea surface, and have captured the imagination
of seafarers since early times, but no marine insects are known that remain submerged
throughout their lives. Some speculations as to why insects are so rare in marine habitats have been based on single parameters, such as the low concentration of calcium
in sea water or the absence of angiosperms. Others have included a complex combination of biological, physical and chemical factors. It is still not understood how such
factors may have operated to exclude an otherwise highly successful taxon of arthropods from the most extensive biome on earth. Possibly all vacant ecological niches had
already been occupied by crustaceans before the insects had evolved on land and were
in a position to colonize the oceans from which their distant ancestors originally
emerged.

To return to an aquatic existence, terrestrial arthropods have to overcome physical,
physiological and ecological problems. For such a return to be possible, there must be
intermediate environments between land and sea, as are provided by the intertidal
zone, estuaries, mud flats, salt marshes and mangrove swamps.

A reduction in weight and consequent increase in surface to volume ratio may be
beneficial to aerial insects, but are an obstacle for species that have to penetrate interfaces between air and water. Hydrofuge and hydrophil regions of the cuticle are important for maintaining orientation in free-swimming and surface-dwelling insects. Some
Collembola, such as *Podura aquatica,* form colonies on the water film. Their feet rest
in depressions of the surface (Fig. 6), but the claws are hydrophil and are drawn into
the water, giving purchase while the hydrophil ventral tube anchors the animal to the
surface. The tarsi of the second and third pairs of legs of water skaters (*Gerris*) are

equipped with hydrophil hairs, while the forelegs are held above the surface and used for seizing prey. The remainder of the insect is covered with a fine velvet-like pile which prevents the water from coming into actual contact with the body. Whirligig beetles (*Gyrinus*) adopt a half and half position. Water is repelled by the smooth and shiny elytra while the lower parts of the body are immersed.

Many marine insects, such as sea-skaters, have become wingless. Some, including chironomid midges, have reduced wings while others still have fully developed wings but are flightless or weak fliers. The latter include shore bugs, beach flies and beetles. Such adaptations perhaps help marine insects from being readily blown away from their proper habitats, as does the flightlessness of montane insects (Chap. 6.3). Water-mites and fresh-water insects that swim actively are equipped with hairy or flattened limbs which act as oars. Aquatic Mymaridae swim through the water by flapping their tiny wings.

Many terrestrial arthropods, subjected to periodic submergence in water, respond by becoming comatose. In a state of akinesis, metabolism in reduced and survival prolonged. The extent to which intertidal species that do not use an air bubble become comatose is not known, although it is probable that some of those which lose their air bubbles survive only by doing so.

6.4.1 Respiratory Adaptations

There are no respiratory adaptations among marine arthropods that are not also found in fresh-water species. Indeed, the number of different types is less in marine insects, as might be expected since they are much fewer than fresh-water species. The success-ful invasion of water by pterygote insects has resulted in six distinct adaptations. These have been summarized by H.E. Hinton (in Cheng 1976) as follows:

a) The ability to make periodic trips to the surface and renew supplies of air. The insects so endowed are among the most numerous of all aquatic forms. They are absent from the intertidal zone, however, chiefly because of the turbulence of the water.

b) Long spiracles or spiracles opening on the ends of long projections. These enable insects to take in air while remaining beneath the surface of the water. A fringe of semi-hydrofuge hairs around the spiracles enables contact to be made between the atmosphere and the air in the tracheae. This adaptation is found among incom-pletely aquatic insects such as *Nepa, Ranatra, Culex* larvae and rat-tailed maggots (*Eristalis*). It is never found in intertidal forms because the depths involved are too great.

c) Conversion of an impermeable cuticle to a permeable one by loss of the wax layer on part or all of the body wall so that cutaneous respiration can take place. This adaptation is found in most aquatic endopterygote larvae.

d) Pointed spiracles, that can be thrust into plant tissues and so tap the intercellular air spaces, are found in some mosquitoes (*Taeniorhynchus*) and the larvae of leaf beetles (*Donacia*). No intertidal insects possess this modification, but some salt marsh and estuarine species do.

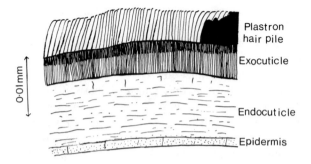

Fig. 57. Section of abdominal sternum of *Aphelocheirus*. The *black area* beneath the plastron hair pile is a region in which air is still present. (After Thorpe 1950)

Plastron hair pile

Exocuticle

Endocuticle

Epidermis

0·01mm

e) In some fresh-water beetles and moths, the gas within the pupal cocoon is in communication with the inter-cellular air spaces of plants.

f) A 'plastron' or permanent physical gill has been evolved independently by a large number of fresh-water insects, and a large proportion of those that have invaded the sea successfully have plastron-bearing spiracular gills in the larval or pupal stages.

6.4.2 Plastron Respiration

A bubble or film of air held under water and in communication with the respiratory system functions both as an air store and as gill. The bubble adhering to the hairs of the water spider *Argyroneta aquatica* provides a fine example. As oxygen is withdrawn from the bubble, the partial pressure of this gas falls, and that of nitrogen increases. Oxygen is more soluble in water than nitrogen is, so the liquid-air interface of the bubble is more permeable to oxygen than to nitrogen. Because of this difference in the solubilities of the two gases, there is a tendency for equilibrium to be restored by oxygen diffusing into the bubble rather than by nitrogen diffusing out of it. Nevertheless, some nitrogen is continually leaking from the bubble which gradually disappears and has to be replenished from the surface. The bubble disappears more rapidly when, under experimental conditions, the gas consists of pure oxygen and there is no nitrogen in it to induce inward diffusion of oxygen from the water. Consequently, when an aquatic insect renews its air bubble, it is more important for it to take up nitrogen than to acquire more oxygen.

In dytiscid water-beetles, air is stored beneath the elytra but, in the great silver water-beetle *Hydrous,* there is not only an air store between the wing cases and the upper surface of the abdomen, but a film of air is held on the under-surface of the body by means of hydrofuge hairs. This film is what confers a silvery appearance to the beetle while it is under water. On rising to the surface, *Hydrous* comes up head first, not tail first as *Dytiscus* does, and inclined to one side. The antenna of that side is used to break the water film by means of its hydrofuge club, so that air passes into the cleft between the head and thorax of the beetle, thus replenishing the air film covering the abdomen.

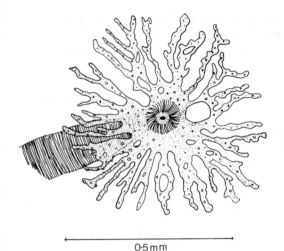

Fig. 58. Spiracular rosette of third abdominal segment of *Aphelocheirus*. (After Thorpe 1950)

0·5mm

A film of air of this kind, held in position by hydrofuge hairs or meshworks of various kinds, is known as a 'plastron'. Some aquatic insects have an incompressible physical gill of this kind, in which the film of air is held by a pile of fine hydrofuge hairs bent over almost at right angles at the tip so that they form a fine hydrophil surface covering a thin film of air (Fig. 57). This arrangement is found in fresh-water beetles of the genera *Haemonia* and *Elmis*, but the finest example is undoubtedly that of the aquatic bug *Aphelocheirus aestivalis*, which can live permanently submerged up to a depth of 7 m. The greater part of the body surface of this remarkable insect is covered with an extremely fine plastron held in position by an epicuticular hair pile having approximately 2,000,000 hairs per mm^2. It is extremely difficult to displace this plastron of air, even by pressures up to four or five atmospheres. Few other aquatic insects are able to live permanently submerged without ever renewing the nitrogen in their plastron.

The abdominal spiracles of *Aphelocheirus* open into a rosette consisting of a number of fine branches radiating beneath the plastron from the tracheal opening, and no closing device is present (Fig. 58). Oxygen is conveyed by the fine branches of the plastron to the tracheae and tracheoles. Carbon dioxide is much more soluble in water than are oxygen or nitrogen and is readily eliminated through the plastron hairs.

Many littoral insects possess plastron-bearing spiracular gills. These are well adapted to the intertidal environment where they are regularly subjected to alternate flooding and drying. In water, the spiracular gill provides a very large surface for diffusion of oxygen, and its structure is such that it does not collapse under the hydrostatic pressures to which it is subjected. Out of water, the relatively enormous surface area of the gill does not involve the insect in excessive water loss over the entire area because the area of the connection between the gill and the internal tissues of the insect is no larger than that of the normal spiracles of terrestrial insects.

Plastron respiration is found in all stages of development. Chorionic plastrons protect the eggs of many species from flooding. Cocoons do not only confer immunity against predators; many pupae that occur in places likely to be flooded are encased in cocoons, the interstices of whose meshworks provide large water-air interfaces which act as plastrons when the cocoons are submerged in water.

Experimental studies on the osmoregulation of marine insects, such as the larvae of the littoral caddis-fly *Philaniscus plebius,* salt marsh mosquitoes, and Corixidae, indicate an ability for osmotic regulation over a wide range of salinities.

Probably more than half the species of insect that have successfully invaded littoral areas have come from land rather than from fresh water. They include Tipulidae, Dolichopodidae, Canaceidae, Hymenoptera and Coleoptera. Notable exceptions are the Chironomidae and Trichoptera. Very few insects invade intertidal areas exposed to wave action. Regular alternate flooding and drying of this environment, rather than its salinity, has made it so inhospitable to insects.

6.5 Cavernicolous Adaptations

The physical conditions within caves tend to vary less than those of the surrounding surface environments. Cave temperature approximates to the mean annual temperature of the region and relative humidity, even in dry passages, rarely falls below 80 per cent. Most cave organisms are found in areas of near saturation, and it is rare to find cavernicolous species in areas that are not visibly damp or wet. A major feature of cave adaptation may be concerned with living with high humidity. The major sources of food are micro-organisms, plant detritus left by flooding, cave cricket eggs and guano, bat guano and the faeces of other mammals.

It is generally assumed that the fragile, delicate morphology of cavernicolous animals results from selection for increased sensory organs on the appendages which, in turn, leads to their lengthening. This is certainly true of interstitial forms which crawl between grains of sand and also of certain groups such as the Orthoptera, Carabidae, Trechinae and Opiliones. But cavernicolous animals, related to stout epigeous types with short limbs, give no impression of slenderness and possess normal appendages. Obligatory troglobitic insects are usually apterous, whilst troglophiles are frequently brachypterous. The disappearance or reduction of wings is associated with a regression of the alary muscles.

The great majority of troglobites are partially or totally colourless, but a few cavernicolous species retain normal pigmentation, even though they live permanently in underground caves. The reduction of the eyes is another phenomenon frequently encountered in cavernicoles. This may take many forms and in all cave-dwelling taxa there are series of reductions from barely perceptible regression to the total disappearance of the eyes and associated structures. Food is generally scarce in caves and a lower metabolic rate allows greater resistance to starvation. Cave-dwelling arthropods have evolved an effective water-excretory mechanism that conserves salts. This involves cuticular reduction, resulting in increased permeability of the integument, which is unimportant in the humid environment.

No single morphological or physiological criterion can, however, be regarded as being uniquely characteristic of cave-dwelling arthropods. All that can be stated is that certain manifestations, in particular depigmentation and the loss of eyes, are more frequent in cavernicolous than in epigean arthropods. The reason why regression of the eyes and loss of pigmentation should occur in uncertain but, in the absence of selection, unfavorable genes inducing these effects would not be eliminated while any metabolic saving produced would be positively beneficial.

6.6 Suspended Animation

Many insects, mites and other arthropods are able to withstand unfavorable conditions in a state of diapause. A few cases are also known in which arthropods do not resist evaporative water loss, but allow their bodies to become almost completely dehydrated. This leads to cryptobiosis, a virtually non-living state in which all metabolic processes have come reversibly to a standstill.

6.6.1 Diapause

The possession of a resting phase of diapause enables many insects, mites and probably some other arthropods to survive the winter in a dormant state which is characterized by enhanced resistance to cold and drought. Diapause is under hormonal control, and is usually induced by decreasing temperature and or photoperiod. It has been the subject of much research. In general, the photoperiodic reaction is independent of the intensity and total energy of the light, provided that this exceeds a minimum threshold value. It is no coincidence that this threshold exceeds the intensity of moonlight.

The selective advantage of diapause in the life cycle of an arthropod is twofold. It not only provides a mechanism for survival when food is scarce and the climate unfavorable, but it synchronizes the development of the individuals in a population so that they emerge as adults and reproduce at the appropriate season. Photoperiod and temperature are the main environmental factors that influence the rate and development of diapause, and serve to maintain it. This implies the possession of (i) receptors that detect the presence of daylight, (ii) a 'biological clock' to measure the length of the photoperiod, and (iii) an effector system to control the metabolic changes necessary for the induction of diapause.

As already indicated, many species show an all-or-none response to daylength, measuring long versus short photoperiods in relation to a relatively well-defined critical photoperiod for the maintenance of diapause. Some show a graded response to the absolute duration of the length of the day, while yet others respond to the direction of change of daylength, according to whether it is increasing or decreasing. The state of diapause is largely a dynamic one and, as the season progresses, decreases in intensity and the animal's responses to the factors that maintain it decrease. Long photoperiods and/or exposure to low temperature may terminate diapause experimentally but, under natural conditions, they usually regulate the rate of its development rather than terminate it actively. Within any one species, however, there is often considerable variation in the duration of diapause.

After diapause has been induced, it is maintained by low temperature and short or decreasing daylength. As autumn proceeds, the response to photoperiod diminishes, and the low temperature threshold rises. In consequence, diapause ends after the winter without the intervention of any specific stimulus. In a few cases, however, environmental changes, such as increasing daylength, food or moisture, serve to end diapause in the spring, and this also synchronizes the development of the various individuals in the population. Summer diapause or aestivation, too, usually requires a specific stimulus for its termination and the same is true among tropical species for the termination of diapause at the appropriate season. After its termination, some species require a specific stimulus, such as the absorption of water, to initiate development. In most cases, however, all that is needed is a temperature within the limits set by the upper and lower thresholds for growth.

In 1934 V.B. Wigglesworth suggested that insects enter diapause on account of a temporary lack of certain hormones. This hypothesis has since been verified experimentally for a wide variety of species. It appears that the effector system controlling the onset of diapause involves glands in the brain. When these are exposed to short photoperiods, the release of hormones is inhibited; when the brain is exposed to longer daylength, their release is promoted.

Two hypotheses as to how insects measure time have been proposed. According to the first, the length of the day or night is measured by an interval timer — an hourglass-type of mechanism. Such a mechanism could be started by dawn and stopped by dusk, or initiated by nightfall and halted at dawn. The second hypothesis, that of E.B. Bünning, is that measurement of the length of the day or night is accomplished by the endogenous daily rhythm or 'biological clock'. Each cycle consists of a photophilic and a scotophilic phase. The endogenous rhythm is set by dawn. If the daily light period is long, the period of illumination extends into the scotophilic part of the cycle, and the organism exhibits a long-day response. If the photoperiod is short, however, the organism exhibits short-day responses. Quite possibly both systems operate in different organisms: clear hourglass effects have been demonstrated in mites and aphids, for example, but this does not preclude the existence of a circadian clock operating simultaneously.

Although both temperature and photoperiod are key environmental triggers in the regulation of diapause, the process is usually mediated by the neuroendocrine system and, in pupae and adult insects, involves withholding the secretion of a stimulatory hormone. In adult insects, the failure to release juvenile hormone inhibits the development of the eggs and, in males, of the accessory glands. Unlike pupal and adult diapause, that of eggs and larvae is induced by the presence of an inhibitory hormone. In some Lepidoptera, larval diapause is limited to the maintenance of high levels of juvenile hormone in the blood. Egg diapause has been examined primarily in the silkworm *Bombyx mori*. It has long been known that injection of an extract of the suboesophageal ganglion of adult females, reared as larvae on a short-day photoperiod, will induce diapause in the eggs of non-treated adults. Subsequent research on this maternal diapause hormone has shown it to be a peptide, but its mode of action is not yet fully understood.

6.6.2 Cryptobiosis

In the cryptobiotic state, all the metabolic processes and chemical reactions responsible for maintenance and growth almost cease, and oxygen consumption is reduced to a negligible amount. The viability of organisms in a state of cryptobiosis decreases with the passage of time, however. This decrease is due to adventitious oxidations and reductions, and is caused by chemical reactions that have nothing to do with metabolism. It can, of course, be greatly accelerated by raising the temperature. For instance, when dehydrated cryptobiotic eggs of the brine shrimp *Artemia salina* are exposed to high temperatures, there is little reduction in the percentage hatching (when subsequently placed in a solution of salt) if the length of exposure has been below a critical period of time. Continued exposure beyond the critical period, however, results in a steep decline in the numbers hatching.

Mst of what is known about cryptobiosis in arthropods is based on H.E. Hinton's work with the chironomid midge *Polypedilum vanderplanki*, whose larvae inhabit shallow rock pools in Nigeria and Uganda. At the beginning of the rainy season, these pools may fill and dry several times. When they dry out, the larvae dry out too, surviving in a state of cryptobiosis. At a relative humidity of 60% their moisture content is 8%, and it falls to 3% at 0% relative humidity. When the pools fill, the larvae rapidly absorb water through their integument until their normal moisture content of 85–90% has been restored and within an hour they may be feeding again. In the dry state, the larvae of *Polypedilum* can withstand exposure to 106°C for 3 hours and 200°C for 5 minutes. They can also tolerate immersion in liquid air (−190°C) or liquid helium (−270°C) and subsequently produce normal adults. A dry larva can be cut in half or into smaller pieces and kept for years. If the pieces are then placed in water they recover for a short time as they take up water, then die as a result of their injuries.

These observations illustrate the sharp distinction between biological age and calendar age in *Polypedilum*. The biological age of a larva may be, say, 2 weeks while its calendar age would be 3 years and 2 weeks if it had been in cryptobiosis for 3 years. Cryptobiosis is an efficient adaptation for coping with the exigencies of a severe and fluctuating environment.

Further Reading

Ahearn GA (1970) The control of water loss in desert tenebrionid beetles. J Exp Biol 53:573–595

Cheng L (ed) (1976) Marine insects. North-Holland, Amsterdam Oxford. American Elsevier, New York

Cloudsley-Thompson JL (1964) Terrestrial animals in dry heat: arthropods. In: Dill DB (ed) Handbook of physiology. Am Physiol Soc, Washington DC 4:451–465

Cloudsley-Thompson JL (1969) The zoology of tropical Africa. Weidenfeld and Nicolson, London

Cloudsley-Thompson JL (1975) Adaptations of Arthropoda to arid environments. Ann Rev Entomol 20:261–283

Cloudsley-Thompson JL (1975) Terrestrial environments. Croom Helm, London

Cloudsley-Thompson JL (ed) (1984) Sahara desert. Pergamon, Oxford New York

Crawford CS (1981) Biology of desert invertebrates. Springer, Berlin Heidelberg New York

Culver DC (1982) Cave life. Evolution and ecology. Harvard Univ Press, Cambridge Mass London

Hadley NF (1972) Desert species and adaptation. Am Sci 60:338–347

Hinton HE (1968) Reversible suspension of metabolism and the origin of life. Proc R Soc (B) 171: 43–57

Lawrence RF (1953) The biology of the cryptic fauna of forests. Balkema, Cape Town Amsterdam

Louw GN, Seely MK (1982) Ecology of desert organisms. Longman, London New York

Mani MS (1962) Introduction to high altitude entomology. Methuen, London

Mani MS (1968) Ecology and biogeography of high altitude insects. Junk, The Hague

Saunders DS (1976) Insect clocks. Pergamon, Oxford New York (International series of pure and applied biology. Zoology division, Vol 54)

Seldon PA (1985) Eurypterid respiration. Phil Trans R Soc Lond B309:219–226

Shaw J, Stobbart RH (1972) The water balance and osmoregulatory physiology of the desert locust (*Schistocerca gregaria*) and other desert and xeric arthropods. Symp Zool Soc Lond 31: 15–38

Tauber MJ, Tauber CA (1976) Insect seasonality: diapause maintenance termination and post diapause development. Ann Rev Entomol 21:81–107

Thorpe WH (1950) Plastron respiration in aquatic insects. Biol Rev 25:344–390

Vandel A (1965) Biospeleology. The biology of cavernicolous animals. Pergamon Press, Oxford London (Translated into English by BE Freeman)

Wallwork JA (1982) Desert soil faunas. Praeger, New York

7 Dispersal and Migration

Animal movements are basically of two types: trivial and migratory. Trivial movements take place within the territory of an individual or habitat of a population, while migratory movements carry the animals that make them away from the area. Most animals disperse from their places of origin at least once in their lives. It is generally assumed that one-time dispersal of this kind must provide some ecological and selective advantages, although these are often difficult to assess. Not infrequently, migration enables arthropods and other animals to escape from temporarily unfavorable conditions, such as winter cold, summer drought, and seasonal absence of food. It is an adaptation of inconstant environments, and an alternative to diapause.

7.1 Migration in Relation to Habitat

7.1.1 The Evolution of Arthropod Migration

Migration is an expensive process because most of the individual migrants fail to reach a suitable new environment and, consequently, die without leaving any progeny. For this reason, the suggestion has been made that the evolutionary importance of migration may lie in the elimination of excess population. If this were true, however, it is difficult but not impossible to see how it might have been selected. Imagine two populations of the same species, one possessing a gene for migration, the other without. Under conditions of hardship or food shortage, most of all of the first would migrate, leaving a small residual population, still containing the migratory gene, yet more fit than the starving, overcrowded and perhaps diseased population from which no migration had taken place. Thus, a group might benefit from the loss of some of its members because there would be fewer individuals to compete with one another when food was scarce. Depending upon the size of the group, this hypothesis might imply acceptance of the highly controversial concept of 'group selection': it therefore cannot be adopted lightly. Another objection lies in the fact that migration often takes place independently of the immediate environment. Indeed, migration should be regarded as an adaptation to adverse circumstances rather than a reaction to them. So, it is not only the beginning but also the end of migratory movement that is important. Its prime evolutionary advantage probably lies in enabling a species to keep pace with changes in the locations of its habitats. Within any taxon, a higher degree of migratory movement is to be found in the species associated with temporary habitats than among those with more permanent ones.

The ecological significance of migration, therefore, resides in enabling animals to spend different parts of their lives in different environments. With the advent of wings, pterygote insects have been particularly well able to exploit this ability, and this probably contributes to their success, as measured by the large numbers of individuals and species in existence (Chap. 9.1).

The migrations of arthropods appear to be stimulated by ecological events at the point of origin more directly than are bird migrations. A likely cause might be the local scarcity of a food plant which, in turn, could be the result of drought or other climatic factors. Neither the proximate factors stimulating migration are known with certainty nor, for that matter, what determines the direction taken (Sect. 7.5).

Physiological and ecological observations on a variety of species are consistent with the following generalizations about arthropod migration: (1) During migration, locomotion is enhanced, while feeding and reproduction are reduced. (b) Migration usually occurs before reproduction. It takes place, therefore, when reproductive values and colonizing abilities are at their maximum. (c) Migrants usually rest in temporary habitats. (d) Both physiological and ecological parameters of migration are modified by environmental factors. Taken together, these criteria establish a basic strategy which can be modified to suit the ecological requirements of individual species. Migration is so complex that it is unlikely to result from the same combinations of physiological and environmental stimuli in different species, or even within different populations of the same species.

7.1.2 Types of Migratory Movement

Locomotory movement is characteristic of most animals: arthropods may move by active walking, swimming, flying and, passively, by phoresy and, as in spiders, by ballooning. According to R.R. Baker (1978), migration can be defined very broadly, as the act of moving from one spatial environment to another. On this definition, all movements, from the enormous flights of monarch butterflies to the foraging flights of bees between flowers, are migrations. The rationale for this is that there is a continuum of movements of various type which are found between these two extremes, and there is evidence that similar biological events and evolutionary rules are common to them all. Nevertheless, the majority of zoologists today would probably reject this definition as being too broad to be useful. It is better to distinguish between trivial and migratory movements as suggested above. An earlier analysis of the phenomenon recognized three types of migratory movement as follows: (1) Emigration. When the pressure exerted by a species on its environment increases as a result of overcrowding, irruptions or emigrations may occur. Examples are afforded by millipedes, woodlice, locusts, termites, dragonflies, aphids, ants, butterflies, moths, and other insects whose populations increase to vast numbers and then invade new areas. (b) Nomadism, the adoption of a wandering life based on the need to find food. (c) True migration involving regular directed movement with a return to the place of origin. With regard to arthropods, however, these distinctions are not valid.

Insect migration may be defined as continued movement in a more or less definite direction, and consists of three phases: (a) Emigration from the place of origin; (b) trans-

Fig. 59. Migration routes of the monarch butterfly in autumn

migration; (c) immigration. Among vertebrates, which are long-lived animals, all three take place within the life-time of an individual but, in insects, migratory flight is halted when reproduction takes place, and is then continued by the offspring. For instance, monarch butterflies (*Danaus plexippus*) migrate from Canada and the Hudsun Bay, where they spend the summer, to Florida, California and Mexico, where the winter is passed in semi-hibernation. With the onset of spring, these southern monarchs migrate north again, towards the Great Lakes. Here they encounter those non-migratory individuals that have hibernated over winter near the Canadian border at the northern limit of the population. Breeding takes place, and up to three generations are passed before the return journey begins again (Fig. 59). It is significant that the individuals which migrate south in the autumn are the same as those which fly north again in spring.

7.2 Migration in Relation to Population Dynamics

The proposition that mobility may provide an alternative to competition in the control of arthropod populations has been hinted at the beginning of this chapter (Sect. 7.1.1). If a population increases solely by births, and declines only by deaths, thus conforming to classical population dynamics, reproduction is limited by mortality caused by environmental factors. If, however, the population assembles through movement and then disperses by migration, explanations of population control can be found in the behaviour of its members. In reality, both processes take place. It seems possible that information about population density, received by the individual through social behaviour, may activate negative feedback — represented in migration, rather than by restraint in reproduction.

It has been recognized for some while that, if animal behaviour could be integrated with population dynamics, the density-dependent negative feedback usually considered

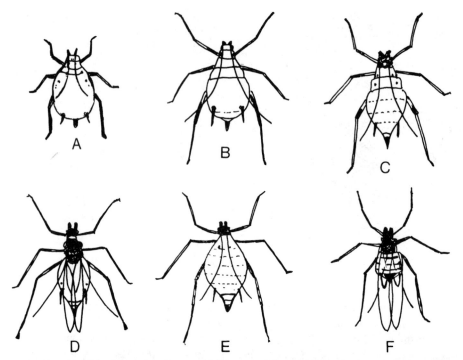

Fig. 60A–F. The polymorphs of an aphid (*Megoura viciae*) identify migratory status as well as sex and reproductive capacity. **A** First generation virginoparous viviparous apterous female: the fundatrix. **B** Second generation virginoparous viviparous apterous female: the fundatrigenia. **C** Apterous virginoparous viviparous female from long-established asexual line. **D** Alate virginoparous viviparous female. **E** Apterous oviparous, sexual female. **F** Alate male. (After Lees 1975)

necessary to maintain control of numbers might come from intrinsic behaviour rather than from extrinsic mortality. The issue at stake would then be as to whether or not fitness would be impaired. Does the emigrant that makes possible the founding of a new population lose fitness in so doing? The suggestion has been made that such a loss of fitness would occur unless the individuals returned at another season. The alternative is that individuals may avoid competition, without loss of fitness, by moving. Fitness can no longer be defined in terms of the number of offspring left to found the next generation because, lacking volition, these offspring would be dispersed at random by environmental forces, and it is well known that spatial randomness is rare among animals.

The great majority of insects are small pterygotes which use their wings mainly for dispersal. Many use them once only when they migrate. Migratory insects are often highly successful. If success is judged by the numbers of individuals, many highly successful species have evolved asexual life-styles secondarily. In their clonal populations, reproductive behaviour is minimal while genotypic and phenotypic variability are clearly segregated. The Aphididae, for example, a highly successful family comprising some 4000 species, illustrates the extent to which behavioral control over a clonal population can be exerted through mechanisms that respond to environmental cues, many of which reflect the pressure of other individuals.

Aphid life histories incorporate alternating cycles of population growth and migration, and usually also of sexual and asexual forms (Fig. 60). The wingless multiplicatory generations are neotaneous, and depend upon the environment for the suppression of wing production, as do the asexual forms for the suppression of the sexual condition. Food, crowding, photoperiod, and temperature, control the formation of wings, the willingness of whose possessors to migrate is activated or inhibited by other factors such as the physical and chemical features of the surfaces from which they take-off or alight, the wind speed, and light intensity.

Each aphid colony originates from a successful migrant that has arrived alone at a new food plant. Here it produces wingless individuals which, in turn, build up the wingless population until a threshold is passed when progressively more of the younger generations develop wings and emigrate so that eventually the colony becomes extinct. Migrants fly to new, more acceptable, habitats free from other individuals, thereby tracking the changing environment. The movement occurs where there is no recognizable sexual motivation and involves no gene selection or recombination.

Sex determination and photoperiodism are interlinked with the control of migration since the suppression of males and oviparous females leads to the virginoparous condition, and hence to further restriction on wing production. The pars intercerebralis of the brain inhibits the production of morphs that reproduce sexually, by responding to the length of the nocturnal dark period interrupting the daylight — an hourglass effect (Chap. 6.6.1). In the absence of inhibition, normal bisexuality is restored. Temperature influences wing production directly, but its side effects are extremely complex.

7.3 Migration Without Flight

Even wingless arthropods migrate. Spiders undertake air-borne journeys by ballooning (Sect. 7.3.2) while, in the case of mites and Pseudoscorpiones, phoresy (Sect. 7.3.3) is frequently employed. Collembola and other very small arthropods are transmitted passively by air currents after being carried inadvertently by wind into the upper atmosphere (Sect. 7.4), while larger ground-living forms migrate on foot.

7.3.1 Mass Migration on the Ground

Mass migrations have been recorded in woodlice and millipedes, as well as in other ground-dwelling arthropods. Since they feed upon decaying plant material which may, occasionally, become exhausted, these animals might well be expected to undergo irruptions or mass emigrations. The earliest recorded migration of millipedes took place in spring, 1876, when great numbers of a variety of species, accompanied also by centipedes, migrated in Transylvania. Two years later, an enormous mass of millipedes actually stopped a train on the Thein Railway in the Hungarian district of Alfold. The millipedes were in such vast numbers that the earth was black with them. The locomotives destroyed them in thousands and were impeded to such an extent that sand had to be strewn before their driving wheels would grip. Similar mass migrations have been recorded throughout the world. As with insects, it seems probable that at least some of these observed mass migrations are just exceptional instances of normal migratory movement and there is, indeed, undoubtedly a wandering tendency in millipedes.

The larvae of certain moths, known as 'processionary caterpillars', are gregarious and construct tents of silk in which they take refuge from incliment weather. They feed at night, travelling to and from the nest in a single-file procession, each larva spinning a thread of silk wherever it goes. The threads of silk left behind by a big procession become quite thick, in some places forming a band 2–3 mm wide; but it is not the threads that guide the caterpillars so much as the tails of the larvae in front. A similar phenomenon is found among larvae of saw-flies. No doubt the conspicuous mass movements of these distasteful, brightly coloured insects warn off potential predators, as do those of the lackey moth *Malacosoma neustria* (Lasiocampidae) (Chap. 8.3.3). Army-worms (Noctuidae) and snake-worms (*Sciara*) are well-known agricultural pests. They are larvae of a number of moth or dipterous species that become migratory when crowded, as locusts do. Migratory caterpillars are intensely gregarious in their mass emigrations and mostly belong to species, such as the noctuid silver Y-moth (*Plusia gamma*), that are well known as migrants when they become adult.

7.3.2 Ballooning

Among the many remarkable behavioral traits of spiders, none has excited greater interest, nor produced more fantastic speculations, than has ballooning. The phenomenon is still under investigation, and there is much yet to be learned about it. Often, during the late summer and autumn months, on quiet sunny days, the air is filled with shining strands and threads of gossamer – the silk produced by innumerable small spiders that have attempted unsuccessfully to fly. Sometimes one sees a meadow carpeted with silk, and a host of little spiderlings vainly spreading their threads.

Aeronautic dispersal of immature spiders takes place mostly during the summer, of adult Linyphiidae in autumn, but there is no season in which some ballooning does not take place. Small spiders climb up grass, fences or sticks, turn to face the wind and, standing on tiptoe, allow air currents to carry silk from their spinnerets. When these exert sufficient pull, the spiders let go and sail away, dragged backwards by their spinnerets. Sometimes, after take-off, a spider will climb rapidly to the middle of its thread, which then sweeps forward and becomes doubled. Less frequently, spiders take off with a forward start, making weak attachments to the support and allowing themselves to be blown upward until the thread snaps (Fig. 61). Ballooning is important in the dispersal of spiders everywhere, but particularly in temperate regions.

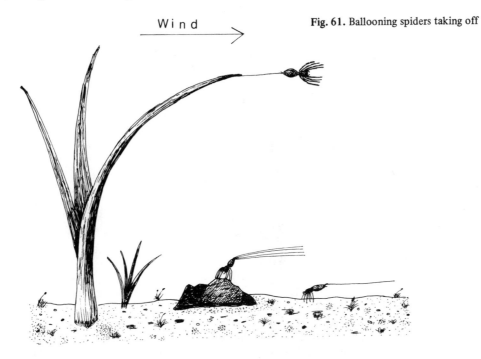

Wind

Fig. 61. Ballooning spiders taking off

7.3.3 Phoresy

During the late Mesozoic era, which gave rise to the angiosperm plants, the evolution
of mites apparently underwent an explosive radiation in conjunction with the evolution and radiation of the insects, and it is not surprising that a seemingly endless
diversity of relationships between these two taxa has arisen. One of them is phoresy
(so named from the Greek *phora:* to carry or bear). Phoresy has been defined as a
phenomenon in which one animal actively seeks out and attaches itself to the outer
surface or another for a limited time, during which feeding and development cease.
This attachment results in movement away from areas unsuited for further development, either of the phoretic individual or its progeny. It is therefore a mechanism
by which dispersal and migration take place.

Phoresy is widespread among arthropods. Small Diptera are frequently transported
by dung beetles to suitable sites for breeding, both for themselves and for the beetles.
Pseudoscorpions and mites are likewise transported by harvestspiders (Fig. 62) and
various types of insects, ants frequently playing the role of porters. When more than
one pseudoscorpion attach themselves to the same fly, they always seem to cling to
legs on opposite sides of the body, so that the insect is not unbalanced. Data on
phoresy accords well with modern definitions of migration as a phase of the depression
of growth-promoting functions, during which the phoretic arthropod is transported
while it shows a special readiness for being moved. Phoresy, not surprisingly, is most
effective where the host gathers within the range of the arthropod. Hence there is
an association between phoresy among mites and the microhabitats between which
the mites need to be carried. Monocultural plant-stands and ecological climaxes are

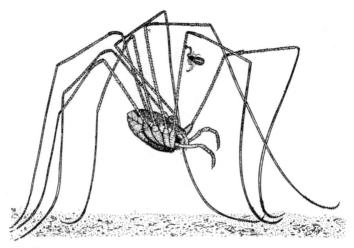

Fig. 62. False-scorpion phoretic on a harvest spider

not much characterized by phoretic associations. Phoresy occurs where habitats are discrete and temporary, and is associated with subclimatic communities. Waiting stages, marked by the depression of growth-promoting functions, occur within the life cycles of phoretic mites. The hypopus is the typical phoretic stage, whose association with the host is an adaptation for survival in extreme environments.

7.4 Meteorological Aspects of Air-Borne Insect Migration

Air-born biological matter, including Collembola, insects, ballooning spiders, mites, and other small arthropods, is regularly carried by the wind from one place to another, sometimes far apart. Once these are air-borne, migration is enforced, although the direction is not well controlled. Long-distance displacement, particularly of small forms, depends very largely upon the speed and direction of air currents. The direction of migration of many small moths, such as the diamond-back (*Plutella maculipennis*) and the small mottled willow moth (*Laphygma exigua*), has been shown to be controlled by meteorological conditions.

7.4.1 Locusts

Apart from Lepidoptera, the most important migratory insects are locusts and other grasshoppers. For many years the mechanism by which swarms of desert locusts (*Schistocerca gregaria*) always seem to arrive in arid localities shortly before rain falls, or immediately after it has fallen, was a mystery. The problem was eventually solved by R.C. Rainey, who related known movements of swarms with concurrent meteorological data. Long-distance migration of swarms takes place high in the air where wind speeds are often greater than the maximum speed of flight. Consequently, locusts

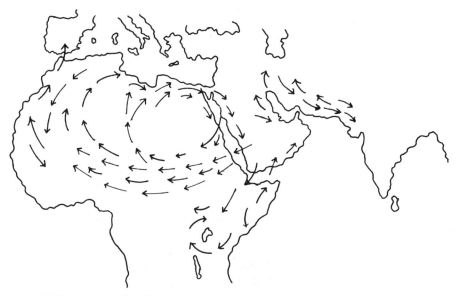

Fig. 63. Migratory circuits of the desert locust

tend to move into areas of low barometric pressure where rain is most likely to fall. The complicated phototactic reactions by which the insects maintain a constant angle with respect to the rays of the sun (Sect. 7.5.3), which attracted so much attention from entomologists in the past, presumably serve merely to keep the swarm together, and exert little effect upon its ultimate displacement. The same remains true if locusts, like butterflies, show preferred compass directions, although this has not yet been established. So, after nearly 2000 years, the observations recorded in the Book of Exodus are confirmed. An east wind brought the plague of locusts to Egypt, and a west wind drove them into the Red Sea!

Nevertheless, locust swarms do describe seasonal circuits which are either circular or cross the same territory (Fig. 63). Although the generation time of *S. gregaria* is too short for individual animals to migrate round the entire circuit, their offspring and descendents manage to do so. For an individual locust and others of its generation, all migration must be emigration; but for the population as a whole the circuit represents true return migration (Sect. 7.1.2).

An interesting method of synchronization of reproduction with environmental conditions is seen in locusts. Although the interval between fledging and oviposition may be as short as 3 weeks, it can be extended to 9 months. In Somalia, for instance, delays of 3 to 5 months are quite common. Yet, even after such an interval, egg laying may occur more or less simultaneously at sites hundreds of kilometres apart. The obvious environmental factor with which this can be correlated is the onset of the rains, but maturation of the gonads begins before the rains actually begin. The signal to which the locusts respond has been shown experimentally to be contact with the terpenoids of aromatic shrubs such as *Commiphora*. These terpenoids are in highest concentration just at bud-burst, and it appears that the desert locusts respond to their scents, which act as a proximate factor stimulating reproduction. The ultimate factor, the one of biological significance, is that eggs should be laid in damp soil and fresh vegetation for the newly-emerged hoppers to feed on after hatching.

7.5 Orientation During Migration

7.5.1 Butterflies

None of the explanations of migration mentioned above accounts for directed flight, a feature long considered to be characteristic of the migration of butterflies. During their evolution, however, many species have become adapted to larval foods that occur in small localities whose distribution changes constantly. These butterflies fly from one site to another, maintaining, it is suggested, a constant angle to the rays of the sun which varies with the ambient temperature. In temperate regions, the peak direction taken by the first and second broods of multi-brood species is towards a lower temperature. It is a function of the geographical temperature gradient experienced during larval development. In the autumn, there is a major change in the direction of flight towards a warmer temperature.

The peak directions of migration of temperate butterflies in western Europe in spring is north-northwest. For example, 42% of all small whites (*Pieris rapae*) fly in this direction and similar results have been obtained from study of the painted lady (*Cynthia cardui*), peacock (*Inachis io*) and large cabbage white (*Pieris brassicae*). In autumn, the peak direction of flight of these species changes to south or south-south-east. Direction is maintained by orienting to the sun's azimuth. Unlike birds, and some other insects such as bees (Sect. 7.5.3) and tropical butterflies, temperate butterflies do not compensate for the movement of the sun. Their angle to the azimuth remains constant at all times, so that the direction of flight changes about 90° from mid-morning to mid-afternoon. Migration takes place during the warmest time of day.

The migration of tropical butterflies is more dramatic than that of temperate species because there is more coordination and they all fly in the same direction. Migrating swarms of butterflies in Africa are sometimes so vast that they stretch for 25 km or more, with stragglers extending considerably further in each direction. Two species of whites (Pieridae) are outstanding migrants south of the Sahara. One of these ranges throughout the continent except in the western tropical forests; the other is widespread in East Africa. The larvae of both feed on spurge, and their food plants are sometimes completely defoliated by many thousands of caterpillars.

7.5.2 Night-Flying Moths

These migrate in much the same way as butterflies do. Most of them take to the air about an hour after sunset and climb quite fast to an altitude of about 400 m. The ceiling is limited by temperature in the subtropical regions of Africa and Australia (where the migration of moths has been studied by radar) to about 1500 m. The rapid ascent is followed by a gradual descent over a period of 1 to 4 hours. A preferred compass direction is probably maintained by time-compensated lunar navigation and, when the moon is below the horizon, by the pattern of the stars (Sect. 7.5.3).

7.5.3 Time-Compensated Celestial Navigation

The ability of honeybees (*Apis mellifera*) to return to a source of food at the same time each day has been known since the beginning of the present century, when it was observed that bees only visited a field of buckwheat in the mornings when the blossoms were secreting nectar. Honeybees not only orient themselves by the direction of the sun, but can also appreciate the plane of polarization of light. Light travels in waves which vibrate transversely; that is, at right angles to the direction of travel. In ordinary light, these vibrations lie in an infinite number of transverse planes but, in polarized light, they are in only one transverse plane. The light reflected from any part of the blue sky is partly polarized, the plane and the degree of polarization depending upon its direction of the sun. The human eye cannot distinguish between ordinary and polarized light, but arthropods can even distinguish the direction of the vibrations, a facility they make use of in their orientation. For this reason, honeybees are in no way disorientated when the sky is obscured by cloud. So long as a part of the sky remains visible, they are able to maintain their sense of direction.

Time-compensated solar navigation depends upon the ability to steer by the sun, at the same time making allowance for its apparent movement across the sky. The position of the sun in the sky can only afford a precise sense of direction if one knows the time. Bees and other arthropods compensate for the sun's movements in this way, measuring time by means of their endogenous 'biological clocks'.

It has long been known that black lawn ants (*Lasius niger*) use the position of the sun — the light compass reaction — to maintain a straight course in territory poor in land marks. If shielded from the direct rays of the sun, but exposed to its reflection in a mirror, they alter their own course as though steering by the sun, which is a much more powerful stimulus than the plane of polarized light in the sky. Earlier work suggested that, if the ants were detained for a few hours in a darkened box and then released, they would continue at the same angle to the sun as before and, therefore, in a different direction. Clearly, like temperate Palaearctic butterflies (Sect. 7.5.1), they do not compensate for the sun's movement by changing their angle of orientation. More recently, however, compensation for the changing azimuth of the sun has been described in wood ants (*Formica rufa*), but only in summer. During March and April, wood ants are apparently unable to compensate for solar movements and show an incorrect compass direction after being imprisoned in the dark. Apparently, compensation for the changing azimuth has to be learned anew each year, after the winter.

Time-compensated sun-compass orientation has also been described in locusts, beetles, pond-skaters (*Velia*), sandhoppers (*Talitrus*), tropical butterflies, wolf-spiders (Lycosidae) and other arthropods. The ability is evidently widespread. Various littoral sandhoppers, for example, are able to return directly towards the sea if moved inland and placed on dry sand. In some experiments, sandhoppers were transported from the west coast of Italy to the Adriatic shore. On release, they persisted in moving westwards, despite the fact that the nearest sea now lay to the east.

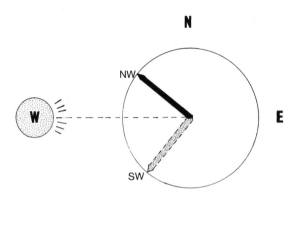

Fig. 64. Effect of phase shifting the 'biological clock' of an animal 6 h anti-clockwise. The sun-compass direction is changed from south-west to north-west

The 'biological clocks' of sandhoppers, like those of other arthropods, are very resistant to environmental changes. When sandhoppers were taken by plane from Italy to South America, they oriented at an angle to the sun that would have been correct had they remained in Europe. The importance to amphipods, which soon become desiccated in dry surroundings, of finding their way directly back to the damp sand of the seashore when accidentally transported inland, or blown there by the wind, is quite obvious. Animals of each population must, however, learn a different direction according to the position of the sea in relation to the beach on which they happen to live. A similar type of orientation has been found among wolf-spiders (*Arctosa perita*) which often live on the banks of rivers and lakes. If placed on the water, these hurry back to the bank in a direction perpendicular to the shore line. The participation of the 'biological clock' or time sense has been demonstrated experimentally by keeping the spiders temporarily in darkness, after which they still head directly for the land. Their sense of direction can, however, be disturbed by phase-shifting their circadian rhythms experimentally. For instance, a 'clock' shift of 6 hours anti-clockwise will result in an animal heading north-west instead of south-west (Fig. 64).

Mechanisms such as those described are, no doubt, involved in the long-distance migration of most terrestrial arthropods. Since all living cells appear to possess intrinsic 'biological clocks' it is not surprising that these should be involved in migration whenever it is beneficial to do so.

Very little is known about lunar orientation in arthropods, but it seems not improbable that nocturnal species navigate by means of time-compensated moon-compass reactions (Sect. 7.5.2). There is some evidence for this in sandhoppers, which tend to move inland at night when the temperature drops and the relative humidity of the air increases, because they appear to be disoriented on moonless nights or when clouds obscure the moon. Woodlice, ants, and various other nocturnal arthropods have likewise been reported to be suddenly disoriented when the moon is hidden.

Stellar navigation appears to be used by moths that migrate at night. Using tethered moths, mainly the large yellow underwing *Noctua pronuba*, experiments have shown that orientation does not occur in a dark room or when the eyes are completely painted over. When the moon or stars are visible, however, the moths show compass orientation. Experiments carried out at different times of night indicate that an individual moth's orientation shifts round as the moon's azimuth moves across the horizon. Like most temperate butterflies with respect to the sun, the yellow underwing does not compensate by means of its 'clock' for movement of this azimuth; but, even when the moon is below the horizon, the tethered yellow underwing shows compass orientation so long as the sky is not overcast. This implies that orientation must take place relative to some star or to the pattern of the stars.

7.6 Migration and Diapause

The physiological and ecological similarities between migration and diapause are considerable. The two phenomena may occur together in adult insects, and migratory flights to and from diapause localities are common. At the same time, both involve suppression of reproduction and survival during periods when environmental conditions become unfavorable. Prior to the pre-diapause flight, there is often an increase in the size of the fat body. This decreases during winter and early spring, at which period there is no ovarian development. Rapid development of the ovaries in late spring in north temperate latitudes coincides with mating and migration to breeding sites.

An example is afforded by the North American milkweed bug *Oncopeltus fasciatus*. This species migrates northwards in the spring, colonizing stands of developing milkweed. The migration proceeds in a series of steps, with each generation advancing a few hundred kilometres further north. When the photoperiod decreases in the autumn, the bugs enter ovarian diapause, reproduction ceases and the adults move southwards, assisted by the prevailing wind, as they mature. Reproduction takes place at various localities in the southern United States, Mexico and the West Indies throughout the winter, for the ovarian diapause is remarkably flexible, although hereditary, and can be shifted with age over a period of a few generations. The stimuli which cause the arrest of migratory flight are frequently associated with the food plant. Feeding may induce degeneration of the wing muscles, thus preventing further migration, or it may induce reproduction, which inhibits prolonged flight. Food-induced reproduction, which suppresses further migration, is seen in locusts as well as in aphids and other Hemiptera.

7.6.1 Physiology

Migratory flight takes many forms in different insect taxa. It varies in the duration of single flights, in their repetition, in the degree to which the migrants can be distracted, and in their orientation. In some taxa, such as locusts and migratory aphids, most of the flight undertaken during adult life is migratory. In others, such as honeybees and non-migratory locusts, most of it is appetitive and trivial. So far, no common physiological cause of migration has been identified and the mechanism clearly differs from one taxon to another. In no species has the lowered flight threshold believed to characterize migratory flight been analyzed in relation to hormonal balance. Not have the biochemical conditions of the flight muscles been related functionally to age or with ovarian immaturity. Many biochemical pathways are involved in the attainment of full flight capacity, and a deficiency in any one of them might cause an individual insect to fail to migrate, if not to fly. Migratory flight is a temporary phenomenon with most insects, and an ontogenetic sequence during development. It is associated with deficiency of ecdysone in the corpora allata, and probably occurs when juvenile hormone and ecdysone are in a particular state of balance. It should, therefore, be seen as an aspect of the development of behaviour patterns in the adult. The correlation with the amount of juvenile hormone in the blood links the physiology of migration with that of diapause which, as we have seen (Chap. 6.6.1), is inhibited in adult insects by the absence of juvenile hormone.

As in the case of diapause, migration can be facultative or obligatory. Obligatory migration is independent of environmental factors although it may be geared to the frequency of the appearance of new habitats or the disappearance of old ones. The most highly evolved condition occurs in arthropods that possess facultative diapause. This is not a response to immediate adversity, but is triggered by some proximal factor which heralds the advent of adverse conditions. In some taxa, such as aphids, species respond facultatively to such factors by producing forms that will, in turn, undergo obligatory migration.

Further Reading

Baker RR (1978) The evolutionary ecology of animal migration. Hodder and Stoughton, London Sydney

Baker RR (ed) (1980) The mystery of migration. Macdonald, London Sydney

Binns ES (1982) Phoresy as migration – some functional aspects of phoresy in mites. Biol Rev 57:571–620

Cloudsley-Thompson JL (1978) Animal migration. Orbis Publishing, London; Instituto Geografica de Agostini, Novara

Cloudsley-Thompson JL (1980) Biological clocks. Their functions in nature. Weidenfeld and Nicolson, London

Dingle H (ed) (1978) Evolution of insect migration and diapause. Springer, Berlin Heidelberg New York

Johnson CG (1969) Migration and dispersal of insects by flight. Methuen, London

Johnson CG (1974) Insect migration: aspects of its physiology. In: Rocksteen M (ed) The physiology of Insecta, Vol 3. Academic Press, New York, pp 279–334

Lees AD (1975) Aphid polymorphism and 'Darwin's demon'. Proc R Entomol Soc Lond (C) 39: 59–64

Schmidt-Koenig K (1975) Migration and homing in animals. Springer, Berlin Heidelberg New York

Southwood TRE (1962) Migration of terrestrial arthropods in relation to habitat. Biol Rev 37: 171–214

Southwood TRE (1981) Ecological aspects of insect migration. In: Aidley DJ (ed) Animal migration. Cambridge University Press, Cambridge London, pp 197–208

Taylor LR, Taylor RAJ (1978) The dynamics of spatial behaviour. In: Abling FJ, Stoddard DM (eds) Population control by social behaviour. Institute of Biology, London, pp 181–212 (Symposia of the Institute of Biology, No 23)

Williams CB (1958) Insect migration. Collins, London (The New Naturalist)

Wynne-Edwards VC (1962) Animal dispersion in relation to social behaviour. Oliver and Boyd, Edinburgh London

8   Defensive Mechanisms

The enemies of terrestrial arthropods, against which defensive mechanisms have been evolved through natural selection, are basically of three kinds:

a) Relatively large vertebrate predators, especially reptiles and birds. Most, if not all, adaptive coloration is directed towards the avoidance of such enemies, while toxins that deter them have been evolved;
b) small invertebrate predators, comprising mainly insects and arachnids;
c) parasites.

The selection exerted by these three influences is different, however, and the adaptations that afford protection from vertebrate predators, for example, are usually ineffective against, say, predatory mites or insect parasitoids.

8.1   Concealment from Vertebrate Predators

Concealment or crypsis is not only the most common, but also the simplest means of escape from the attentions of predatory enemies. The ways in which concealment and camouflage are achieved in nature have been analyzed by H.B. Cott (1940). They are as follows: (a) General colour resemblance to the environment. This may vary with season or life history; (b) obliterative countershading. Crypsis is achieved by coloration that counters the effects of shadow; (c) disruptive coloration, which breaks up the outline of the animal. This involves concealment of the appendages and of the eyes; (d) concealment of shadow by shape and posture. Among larger animals, concealment may have an offensive function, as when a leopard stalks its prey, but the animals upon which predatory arthropods feed are usually too small for aggressive concealment to be of significance.

8.1.1   Cryptic Coloration

Most cryptic arthropods tend to resemble the background colour of their environment. Thus, the insect fauna of tropical rainforests is predominantly green, while desert arthropods are usually a shade of brown – or black. The function of black coloration in desert animals is aposematic: it advertises and makes its possessors conspicuous (Sect. 8.2.1). The colours of insects and spiders, too, are often well adapted to the habitats of their possessors. Species that live on bark are usually brown, those that dwell on sandy ground are sand-coloured, while those that live in grass are green.

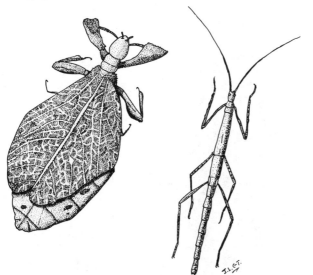

Fig. 65. A leaf-insect and a stick-insect, both members of the same insect order (Phasmida). (Cloudsley-Thompson 1980)

Obliterative countershading can be demonstrated among caterpillars whose under-surfaces are usually pale, while the upper surfaces are darker. In the larva of the eyed hawk-moth *Smerinthus ocellatus,* which normally rests upside down beneath a twig, the colours are reversed and the dorsal surface is pale. Disruptive colour patterns which draw attention away from the shape of the arthropod that possesses them are seen in innumerable species of moths, grasshoppers, spiders, Opiliones, and so on. Finally, animals with cryptic coloration invariably adopt postures that enhance the conceal-ment afforded by their camouflage.

8.1.2 Protective Resemblance

Mimicry by an animal of an inanimate object such as a stick, a stone, a leaf or the droppings of a bird, is an effective escape from predation. Whether this should be regarded as a special case of batesian mimicry (Sect. 8.2.2) is a matter of dispute. It might equally well be considered within the definitions of camouflage and crypsis, depending upon the way in which mimicry itself is defined. The dispute may, there-fore, be one of semantics and little more than a quibble about the meaning of words. Furthermore, an arthropod that, from a distance could be regarded as being merely cryptic may, at close quarters, show protective resemblance to a stick or a piece of lichen growing on bark (Sect. 8.2.4).

Protective resemblance is a widespread defensive device among arthropods. Familiar examples include stick-insects and leaf-insects (Phasmida) (Fig. 65), the mantids and geometrid caterpillars that mimic twigs and sticks. Many butterflies rest with their wings closed to that they look like dead leaves. Tenebrionid beetles of the genus *Cossyphus,* found in tropical Africa, Asia and the Mediterranean regions of Europe, like members of the Australian genus *Helaeus,* mimic winged seeds (Fig. 66). The wing cases are greatly expanded, giving the beetles a flattened, oval shape, while the body is

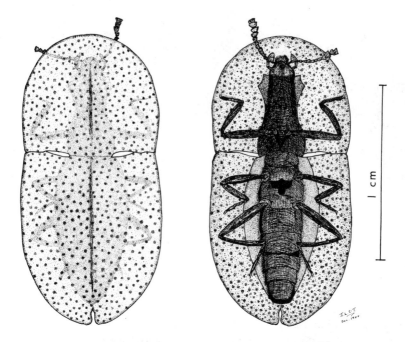

Fig. 66. An East African tenebrionid beetle (*Cossyphus*) showing resemblance to a winged seed. *Left* dorsal view; *right* ventral view. (Cloudsley-Thompson 1980)

only slightly thickened on its ventral side. Many Saharan grasshoppers and mantids, when motionless, look exactly like the numerous small stones which litter the desert in the regions where they live. Their habit of running fast and stopping abruptly makes it almost impossible to discover the exact spot in which they have come to rest (Sect. 8.2.3).

Several unrelated insects throughout the world are disguised as excrement, but the deceptive appearance may be brought about in entirely different ways. Caterpillars of several species of swallowtail butterfly mimic bird droppings and, while they are small, rest conspicuously on the upper surfaces of leaves. When they grow too large to look like bird droppings, however, they become green and cryptic, and remain on the stems and undersides of the leaves. Small caterpillars sometimes mass together, the whole group simulating the droppings of a bird. The twisted masses of egg-sacs of an Indian scale-insect achieve the same effect. Certain tropical beetles likewise mimic the excrement of birds, lizards, or even of caterpillars. A particularly striking example of this type of disguise is provided by a Javanese crab-spider, which constructs on the surface of some prominent dark green leaf an irregularly-shaped film of web. On this the spider lies on its back, with its black legs crossed over its body, exposing the white ventral surface of the abdomen. The black and white together look like the central dark portions of the excreta, while the thin web represents the marginal watery portion becoming dry. Although this disguise is sometimes cited as an example of alluring coloration – many butterflies regularly feed on bird droppings – it is actually an anti-predator device. The sharp eyes of birds are better able to distinguish between excrement and a spider than are the compound eyes of a butterfly. Furthermore, since insect vision extends further into the ultraviolet than does that of vertebrates, it is improbable that the resemblance to excrement would be so perfect when viewed by a butterfly rather than by a bird or lizard.

8.1.3 Disguise with the Aid of Adventitious Materials

Many arthropods obtain protection from predatory enemies by means of adventitious material that adheres to their bodies. In this way, they not only disguise themselves, but may also make themselves unpalatable. The device is found in several families of insects belonging to a number of distinct orders, and may occur in larvae, pupae, adults, or all three. Caddis larvae provide a well-known example. Their cases of sand grains, twigs and other vegetation not only render them cryptic, but probably make them distasteful to predators as well. The larvae of bag-worm moths (Psychidae) likewise inhabit cases of silk, covered with fragments of vegetable matter. They do not construct a new case at each ecdysis, however, but enlarge the original one as they grow. They anchor it to a leaf or to the bark of a tree and close up the opening when they moult. The females remain inside their larval cases after pupation, but the males are swift fliers.

Many lepidopterous larvae live in cases constructed from the fragments of the substance upon which they feed. The larval cases of clothes' moths (*Tineola*) are a well-known example. A more macabre example of the use of adventitious material for disguise is provided by African assassin-bugs (*Africanthus*), which fasten the corpses of their victims onto their backs, after first sucking them dry. Other reduviid bugs cover themselves with sand or dust, while predatory neuropteran larvae of the genus *Hemerobius,* which live among moss on the bark of rain-forest trees in South America, cover themselves with hillocks of moss or lichen. Beetle larvae of various species cover their bodies with their cast integuments, excreta, or soil, and oribatid mites do the same. Although lace-wing larvae sometimes carry on their backs loose packets of debris, including the remains of insect prey, camouflage appears not to be the principal function of this adventitious covering. Ants and other insects which attempt to bite the larva find the packets moved like shields in their direction, and may end up with only distasteful material in their jaws.

In most instances of protective resemblance, the animal looks like some object that is normally avoided or ignored by predators – a stick, a dead leaf, bird droppings or debris. Sometimes, however, the animal does not change its appearance but, instead, alters its surroundings so that it blends into them, or else makes a number of dummies of itself so that its chances of escaping attack are increased. Many fine examples are provided by orb-web spiders which place pellets containing dead insects, vegetation and so on, on their webs. Many build bands of silk, known as 'stabilimenta' into their snares; these radiate outwards from the hub. To the human eye, they are very conspicuous against a dark background of tropical vegetation, but they draw attention

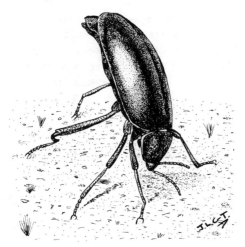

Fig. 67. Darkling beetle (*Eleodes*) in defensive posture. (Cloudsley-Thompson 1980 after W. Wickler)

away from the body of the spider. The zigzag bands appear to be continuous with its body as it rests in the centre of its web, four legs pointing forwards and four backwards. An incidental function of the stabilimentum may lie in making the web conspicuous so that flying birds do not inadvertently destroy it.

8.2 Advertisement with Respect to Vertebrate Predators

Animals that are formidable and well armed, venomous, or distasteful, tend to be avoided by the majority of potential predators. It pays them to be conspicuous, not only because they are easily recognized, but because it is easier for predators to learn to avoid them. Conspicuousness is attained in nature by the bold, striking warning colours of aposematic animals. This conspicuousness is usually enhanced by their free exposure, sluggish behaviour, gregarious and diurnal habits.

8.2.1 Aposematic Coloration

Arthropods with aposematic coloration include wasps whose poisonous stings are associated with conspicuous black and yellow colours, and cinnabar larvae (*Callimorpha jacobaeae*). The larvae of sawflies (Symphyta) are distasteful to predators and are also conspicuously black and yellow. Distasteful desert insects, such as tenebrionid and scarabaeid beetles, are usually completely black. (Against a sandy background, black is presumably even more conspicuous than a combination of black and yellow). The North American darkling beetle *Eleodes* stands on its head when disturbed, and sprays a defensive secretion at the attacker from glands in its abdomen (Fig. 67).

Conspicuousness can be achieved by warning sounds as well as by aposematic colours. The buzzing of wasps and bees, probably caused by vibration of the wings and thoracic musculature and by expulsion of air through the spiracles, may be interpreted as threat to potential attackers. So may the stridulations of Solifugae and some scorpions for, like snakes, these animals are deaf to the sounds they themselves produce.

8.2.2 Mimicry

Distasteful, poisonous, and otherwise formidable animals with aposematic coloration are frequently mimicked by harmless species which resemble them sufficiently closely to be mistaken for them, and therefore avoided by predators. Two kinds of mimicry are often distinguished: batesian mimicry, described above, and mullerian mimicry in which two or more distasteful or harmful species resemble one another. Predators have to learn by experience those species that are best avoided so that, when several distasteful species resemble one another, fewer individuals are damaged or killed as the predators learn to leave them alone. There must therefore be a continual selective pressure for batesian mimicry to become mullerian. Moreover, harmful models must be very much more numerous than their batesian mimics. If this were not the case, there would be a strong chance that naive predators would eat harmless mimics before experiencing their unpleasant models, and thus learn the wrong lesson!

Although it is customary to separate batesian from mullerian mimicry, the significance between them can only be interpreted in relation to the response of the animal that receives the signal from the mimic. Since the tastes of predatory animals vary considerably, mimicry can be effective in the case of one species of predator, and useless in that of another. Furthermore, feeding habits may also vary according to season and the relative abundance of other food. When food is scarce, predators may devour mimics that they would otherwise avoid. With respect to a particular predator, therefore, there may be a switch between batesian and mullerian mimicry according to the season of the year.

An example of the complexity of the relationship between batesian and mullerian mimicry is provided by the milkweed butterfly *Danaus chrysippus,* an aposematic species widely dispersed throughout Africa and the Oriental region. Its large size and reddish wings, with black spots on their margins, make it very conspicuous. It does not synthesize its own toxins, like many insects, but is poisonous because its larvae feed on plants of the milkweed family (Asclepiadaceae). From these they obtain heart poisons (or cardiac glycosides), which exert a strongly emetic action when eaten by birds. These poisons from the plant persist in the pupae and adult butterflies and may be more or less concentrated depending upon the larval food. *D. chrysippus* is mimicked by no less than 33 species of butterflies and moths – either as batesian or as mullerian mimics – in Africa; and five in the Oriental region (where many other species of *Danaus* are available as models).

The related monarch butterfly (*D. plexippus*) of North America (Chap. 7.1.2) likewise obtains poisons from the milkweed plants on which its larvae feed. Again, some individuals are much more toxic than others, depending upon the species of larval food plant. Non-poisonous individuals have been reared experimentally by feeding larvae upon a non-toxic species of milkweed. It is sometimes possible to persuade bluejays to feed upon such non-emetic monarch butterflies. When this occurs, the bluejays peck the butterflies and swallow the pieces gingerly, often regurgitating before finally

Fig. 68. A spider carrying a dead ant and, at the same time, mimicking an ant doing so. (Cloudsley-Thompson 1980 after R.W.G. Hingston)

swallowing them. Obviously they remember the unfortunate effects of cating such insects in the past.

On account of their poisonous stings, ants, bees and wasps are everywhere respected by predators and, consequently, serve as models for batesian mimics from insects as dissimilar as grasshoppers, mantids, bugs, moths, flies and beetles. Ant-mimics occur in a number of spider families including the Clubionidae, Gnaphosidae, crab-spiders (Thomisidae), and jumping-spiders (Salticidae). In some cases, several different species of ants may be mimicked by the same species of spider in its various instars and sexes. Some species of the salticid genes *Myrmarachne* (Fig. 68), of which over 160 have been described, may mimic several different species and castes of ants simultaneously. Others merely have a generalized ant-like appearance. The models of salticid spiders are not limited to ants. Mutillid wasps, sometimes known as velvet-ants, are not infrequently mimicked, while Oriental species of the genera *Marengo* and *Cheliferoides* resemble pseudoscorpions, with which they live.

Most ant-mimics produce the appearance of a narrow waist by means of white patches on either side of a dark-coloured body, leaving only a thin central line. An ant-like appearance is enhanced by running about in a jerky manner. Some spiders carry over their backs the empty dried skeletons of real ants in such a way as to hide their own bodies completely from view. In one bizarre instance, the spider mimics an ant in reverse. The abdomen of the Bornean genus *Orsima* is strongly constricted, while the posterior portion is wide and shaped like a head. The spinnerets are long, giving the illusion of antennae and mouth parts. The resemblance to an ant is increased by the movements and posture of the spider, which often stands with its posterior 'head' raised. In another case of reverse mimicry, also reported from Borneo, dark spots on the orange abdomen of *Amyciaea* give the appearance of the head of a treee-dwelling ant (*Oecophylla*) with its dark compound eyes (Fig. 69). In its movements, however, the spider is ant-like in the ordinary fashion: it behaves as though its front end were the ant's head and does not usually move backwards. Consequently, the ant mimicry is effective mainly when the spider is at rest. Even so, eye spots at the wrong end of the body probably allow a greater opportunity of escape when the spider is attacked by an enemy.

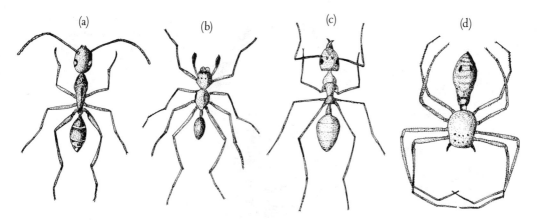

Fig. 69a–d. Spiders that mimic ants. **a** *Camponotus* and its mimic; **b** *Myrmarachne;* **c** *Oecophylla* with its mimic; **d** *Amyciaea* (which mimics in reverse). (Cloudsley-Thompson 1980 after L. Berland)

8.2.3 Thanatosis and Deflection of Attack

By means of feigning death, or thanatosis, the attack of a predator may be inhibited because the prey fails to release the killing response of its enemy. This is believed to be the explanation of the fact that some beetles, bugs, grasshoppers, stick-insects, and mantids become inert when attacked, either with the legs extended, as though their possessors were dead, or else withdrawn close to the body so that they cannot easily be bitten.

Small eyespots on the wings of many butterflies serve to deflect the attacks of predatory birds to non-essential parts where they do little harm. This has been confirmed many times by the fact that triangular beak marks tend to be concentrated near the edges of the wings of species that bear eyespots and not near the vulnerable body. Colours which momentarily flash into conspicuousness when an animal moves, but are normally concealed, help to deflect the attacks of enemies by misrepresenting the whereabouts of the prey, as in the case of the grasshoppers and mantids which look like stones (Sect. 8.1.2). The sudden disappearance of conspicuous wing colours, combined with suspension of movement, tends to render the owner's whereabouts very difficult to detect after it has alighted. This type of defence mechanism is exhibited widely amongst Lepidoptera, Orthoptera, Phasmida, cicadas and other Hemiptera.

Autotomy, or self-amputation of part of the body, is an effective defence by deflecting attack whilst the prey makes its escape. Many arthropods autotomize their limbs in defence: these are regenerated at the next and subsequent moults. Although larger tropical scolopendromorph and scutigeromorph centipedes can deliver venomous bites (Sect. 8.3.1), they are eaten with relish by many lizards, snakes, birds and mammals that are quick enough to catch them. When disturbed, however, the posterior and legs of scolopendromorphs and all scutigeromorph legs can be autotomized. A vigorous contraction and relaxation of the muscles of the severed limb is accompanied by creaking sounds. While the detached limb distracts the enemy, its owner is able to slip away unobserved.

Fig. 70. Pupae of two species of moth (*Spalgis*) which mimic the faces of monkeys. (Cloudsley-Thompson 1980)

8.2.4 Deimatic Behaviour

Many cryptic insects respond to discovery by adopting an intimidating posture. For instance, some tropical stick-insects, mantids and moths have dramatic anti-predator displays that are clearly bluff rather than genuine warning of noxious qualities. Such deimatic behaviour (so named from the Greek word meaning to frighten) consists of suddenly spreading the wings to reveal bright colours or eye spots on the hind wings. These deimatic eye spots do not have the same function as those that deflect predatory attack (Sect. 8.2.3). They give to their owner the appearance of a considerable increase in size. Some lepidopterous larvae display by rearing their bodies so that they look like aggressive snakes, while the pupa of a Burmese moth (*Tonica*) bears a very close likeness to the head of a wolf-snake, a bird-eating reptile widely distributed in Oriental regions. This type of display might equally well be interpreted as a form of mimicry.

Quite a number of arthropods, especially in the tropics, have several lines of defence (Sect. 8.1.2). From afar, the pupae of butterflies of the genus *Spalgis* resemble bird droppings. When seen close up, however, they show an extraordinary resemblance to the faces of monkeys (Fig. 70). One Oriental species is said to look like the common macaque, while a related African species bears some similarity to the face of a chimpanzee. The difficulty in accepting the explanation of disguise or mimicry in such cases lies in the great difference in size between a pupa and the true face of a monkey. This has, however, been explained in terms of hunting by the method of rapid peering. Because their binocular field is narrow, many insectivorous birds gain a perceptance of solidity and distance by evoking parallax. They peer at potential prey from several different angles in rapid succession. From time to time, therefore, a bird will suddenly have a close-up frontal view of one of these pupae, and may not wait to peer at it from another angle!

A similar difficulty lies in explaining the crocodile-like appearance of certain Brazilian fulgorid bugs commonly known as lantern-flies. In many genera, the head is drawn out to form a huge projection that was, at one time, erroneously believed to be luminous. The alligator-bug, in particular, may well gain protection from its appear-

Fig. 71. Head of the alligator-bug (*Fulgora*) showing resemblance to a spectacled caiman (*Caiman*) (*below*). Not to scale. (Cloudsley-Thompson 1980)

ance. It is not inconceivable that, just as a human may recoil in horror from a piece of rope that he mistakes for a snake, so might a monkey that gets a fleeting impression of an alligator be sufficiently startled to allow the creature to escape. Even if the bug is not mistaken for an alligator or caiman, the appearance of a row of formidable teeth may, in itself, prove to be a deterrent to further investigation. Since other large bugs, such as cicadas, are relished by monkeys, there may well have been heavy selection in favour of an alligator-like appearance (Fig. 71).

Whether devices such as these should be regarded as protective resemblance, mimicry, or deimatic behaviour is a matter of opinion and of interpretation, but it does seem possible that somewhat more than chance likenesses may be involved.

8.3 Chemical Defences

It has been calculated that, of at least 80,000 species of poisonous arthropods in existence, no fewer than 50,000 are insects. Some species produce toxic venoms, others distasteful, or repugnatorial fluids. Numerous pharmacologically active substances are secreted by different species, in addition to the compounds that are used as sex attractants, alarm pheromones and so on. For details about their chemistry, the reader is referred to the references listed at the end of this chapter. Without the backing of noxious deterrents, aposematism would never have evolved.

8.3.1 Venoms

Unlike the secretions of repugnatorial glands discussed below (Sect. 8.3.2), venoms are injected by bites or stings, and are usually ineffectual when merely applied to the skin of the foe or taken into its alimentary canal. They probably have an origin in the needs of nutrition, and their defensive properties are secondary. Thus, the stings of aculeate Hymenoptera evolved from ovipositors used as weapons for killing prey, or for boring holes into leaves and wood in which to lay eggs. Their defensive function was acquired afterwards.

The best known stinging aculeates are representatives of the genera *Apis* (honey-bees), *Vespa* (wasps and hornets). Wood ants (*Formica*) possess a vestigial sting which is no longer suitable for stinging, but concentrations of formic acid up to 70% are secreted by the venom glands and prove an effective deterrent when sprayed at any animal that disturbs the nest. Other ants, such as the fire ant *Selenopsis geminata* of tropical America, can sting severely. When an aculeate hymenopteran stings, the pointed aculeus is pushed out from the sting chamber and inserted into the skin of the enemy. Then, by boring movements of the lancets, it penetrates deeply and the venom is injected. The venom glands are derived phylogenetically from ectodermal glands of the ovipositor. Several kinds occur, of which the venom gland s. str. and the Dufour or accessory gland are those most frequently present (Fig. 72). The function of the latter is imperfectly understood. The venom contains proteins and enzymes and causes the release of histamines. In wasps and ants the venom gland also produces alarm pheromones, but these are secreted by the Dufour glands of certain ants (Campono-tinae and Myrmecinae). The gland that produces the alarm substances of honeybees has not yet been identified. In addition to their toxic effects, the venoms of Hymenop-tera are distasteful to some birds, which will not eat these insects, and their odours sometimes repel other insects.

The sting of a honeybee possess a preformed breaking point. It also has strong barbs which anchor it to the skin of the victim. In consequence, the entire sting apparatus, including its nerve ganglion, is ripped out after insertion, and continues working so that every drop of venom can be injected into the body of the enemy. Bee stings have been evolved for use against mammalian enemies. The attack of a large vertebrate endangers the survival of the entire hive, so the loss of a few dozen workers in warding off enemy attack is not significant to the colony.

Many insects possess salivary pumps for injecting poison into the bodies of their prey and some, such as assassin-bugs (Reduviidae) (Fig. 73), use the same mechanism in defence, as well as in spraying their enemies with poison. They will readily jab their proboscis into any animal, including man, and the bite can be extremely painful. It is believed that the habit of feeding on mammalian blood may have been acquired, in the first instance, from the use of the proboscis in defence. The pain has become reduced with the development of regular blood sucking, and species that normally feed on mammalian blood can do so without disturbing their victims, although the bites cause considerable irritation later.

Both scolopendromorph centipedes and theraphosid (tarantula) spiders are able to defend themselves with poisonous bites but, in the case of tarantulas, the urticating hairs on the abdomen are probably even more effective in deterring predatory attack.

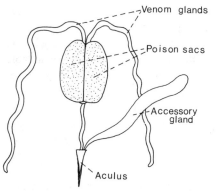

Fig. 72. Poison glands of a wasp. (After L. Bordas in Richards and Davies 1977)

Fig. 73. Assassin-bug (*Triatoma*). (J.L. Cloudsley-Thompson)

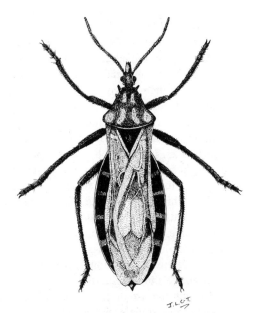

The exceptionally severe effect on mammals of the venom of black widow and show-button spiders (*Latrodectus*) is probably coincidental. Few spiders have fangs sufficiently long to penetrate the skin of a vertebrate. Consequently spiders tend to rely upon concealment, disguise or mimicry to avoid predation, rather than on their venoms.

The stings of scorpions probably evolved in the first instance as a means of subjugating prey, but they are used secondarily in defence. Nevertheless, baboons and other monkeys become adept at catching scorpions without getting stung, tearing off the tail and greedily devouring the rest of the body. No defence can be one hundred per cent effective: despite their stings, scorpions are extremely vulnerable to vertebrate enemies on account of their large size and slow locomotion. The long beak of a stork could easily snap one up, and an antelope might trample a scorpion underfoot, just as mule-deer will stamp on a rattlesnake. Consequently, scorpions are forced to be nocturnal in habit, and many species come out less frequently in bright moonlight than on dark nights. In the USA, the main predators of scorpions, apart from other scorpions, are owls and lizards.

Arthropod venoms contain a wide variety of compounds, some of which constitute unique natural products. Only a few of them have been examined critically, however, and it is premature to make wide generalizations about them.

8.3.2 Repugnatorial Fluids

Distasteful chemicals in general tend to be more effective deterrents to predation than are venomous stings and bites. They are of two kinds: those that are elaborated by special glands, and those that are contained in the blood, alimentary canal, or elsewhere in the body. Mention has already been made of the defensive spray of darkling beetles (*Eleodes*) (Sect. 8.2.1): these are an example of the first. The poison of the milkweed butterfly (*Danaus*) (Sect. 8.2.2) is an example of the second.

Although millipedes are, on occasion, eaten by a variety of animals, including toads and birds especially, to the majority of would-be predators they are unpalatable because of their tough integuments and the irritant exudate secreted by their repugnatorial glands. In most cases, the secretion is exuded fairly slowly but, in some tropical forms, it can be discharged as a fine spray. The fluid contains iodine, quinine and hydrocyanic acid. It has a strong caustic action and is dangerous to the eyes. In addition to their poisonous bites, centipedes produce irritating exudates which are sometimes phosphorescent. They are effective not only against vertebrate enemies but act as deterrents to attack by ants and other insects.

The whip-scorpions (*Mastigoproctus*) of Mexico and the southern United States have glands that open on a revolving knob at the rear of the body from which a vinegar-like acid secretion can be exuded. Numerous insects, including cockroaches, some stick-insects, earwigs, assassin-bugs, and many kinds of beetles are able to discharge defensive secretions when attacked. Most of these are effective against both invertebrate and vertebrate enemies alike.

In addition to the secretions of repugnatorial glands, many arthropods, such as beetles, possess defensive chemicals in their body tissues and blood. Reflex bleeding from the joints of the legs may be stimulated by attack. The blood provides effective protection against ants which, no doubt, have been a major agent in the evolution of this defensive mechanism. Reflex bleeding from blister-beetles (Meloidae) is an effective deterrent to predaceous arthropods as well as to vertebrates, because the blood contains cantharidin, an extremely powerful vesicant. Insects with protective chemicals in their blood sometimes possess separate repugnatorial glands as well, which contain the same toxin as that present in the blood. Some insects defaecate in defence, while grasshoppers regurgitate a fluid from their crops which is toxic to mammals: it is an irritant to the eyes and induces vomiting when swallowed. It also disperses attacking ants. Such chemicals are usually sequestered from the plants upon which the insects feed.

8.3.3 Urticating Hairs

The urticating hairs of theraphosid spiders have already been mentioned. Caterpillars of various families of Lepidoptera are also protected by barbed, urticating hairs which break off when they enter the skin of a predator. In some cases their effect is only mechanical but, in many others, the hairs are poisonous also. The principal families involved are Nymphalidae, Megalopygidae, Eucleidae, Thaumetopoeidae, Limantriidae, Arctiidae, Noctuidae, Saturnidae and Hemileucidae. Caterpillars with urticating hairs have a world-wide distribution. By clustering together, caterpillars of the lackey moth, covered with urticating hairs, present an intimidating spectacle to any predator.

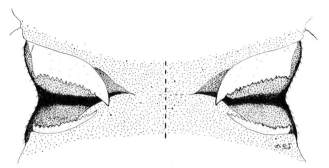

Fig. 74. Second abdominal gin-traps of a beetle pupa (*Alphitobius*). (Cloudsley-Thompson 1980 modified after Hinton 1955)

8.4 Defences Against Small Invertebrate Enemies

As already pointed out (Sect. 8.3.2) many repugnatorial secretions are effective against both vertebrate and invertebrate enemies, but those associated with aposematic coloration must be directed primarily against larger vertebrate predators. The defenses against smaller invertebrate predators are often quite different. Most of them are found amongst insect pupae which tend to be immobile and unable to escape from the attacks of marauding insects, mites and so on. They have been analyzed in detail by H.E. Hinton (1953). They include both passive devices, such as pensile cocoons suspended by a silken cord so that they swing away when pecked by birds, and ants cannot reach them, and non-passive devices. Among the latter are pupal mandibles, gin-traps and ant-attractant glands. The ants that are attracted protect the pupae.

The organs known as 'gin-traps', are widely distributed among the Coleoptera, and also occur in some Lepidoptera. They are confined to the abdomen, the only part of a chrysalis or pupa that can be moved freely, and are essentially a means of defence against mites and other animals such smaller than the pupae. The circumferences of insect abdominal segments are less where they are joined to the adjacent segments in front and behind than in the middle. In consequene, the region near the posterior margin of one segment and that near the anterior edge of the segment immediately behind is always depressed. The evolution of a gin-trap needs no more alteration to the existing structure than a slight deepening of this depression, a hardening of the cuticle at its edges, and their extension into a spiny, keel-shaped blade (Fig. 74). At rest, the abdomen curves downwards so that the jaws of each dorsal gin-trap are held widely open. When the pupa is stimulated, however, the abdomen is bent sharply upwards, and the segments are partly telescoped into each other so that the jaws of the trap meet and overlap. Gin-traps on the sides of the body, on the other hand, are snapped shut when the abdomen makes a lateral or eliptical movement. Gin-traps remain closed only for a fraction of a second, but pinch for long enough to cause an attacker to leave the pupa alone. They do not trap it for any length of time when its struggles might well cause damage to the pupa.

8.5 Avoidance of Parasites

Almost half of known insect species are parasites or parasitoids. To some extent the attacks of these enemies are repulsed physically. Spiders will fight desperately with spider-hunting wasps, and female mantids actively defend their eggs from hymenopteran egg parasites. The egg masses are enveloped with a frothy secretion which rapidly hardens into a firm covering. The chalcid *Rielia*, however, is phoretic on mantids. It is able to oviposit in their egg masses while the covering is still fluid. Hence, it needs to be on the spot when the eggs are laid. Lepidopterous pupae wriggle their abdomens violently when attacked by Ichneumonidae and gin-traps (Sect. 8.4) are often effective in such circumstances. Some rely for defence upon the smoothness and toughness of their cuticles, and are relatively safe from parsitoids after they have hardened. Aphids are often protected from parasitic attack by the ants with which they are symbiotic.

Even after oviposition has taken place, the eggs of parasitoids do not always survive within the bodies of their hosts. The most common function of the haemocytes in the haemolymph of insects and other arthropods is the phagocytosis of foreign particles, micro-organisms such as protozoan and fungal parasites, and so on. Parasites which are too large to be contained by phagocytosis are encapsulated by haemocytes which congregate round them and become flattened. In time the capsule shrinks and is transformed into non-living connective tissue. Encapsulation usually occurs if the parasite is in an unusual host: parasites do not normally evoke encapsulation in their habitual hosts. Some parasitic Hymenoptera resist encapsulation by making vigorous movements, and older tachinid larvae have a respiratory funnel connected to the tracheal system of the host so that their respiration is not impaired. Encysted metacercariae have such low requirements of oxygen that they are not affected by encapsulation.

Further Reading

Blum MS (1981) Chemical defences of arthropods. Academic Press, New York London
Bucherl W, Buckley EE (1971) Venomous animals and their venoms. Volume III Venomous invertebrates. Academic Press, New York London
Cloudsley-Thompson JL (1980) Tooth and claw. Defensive strategies in the animal world. Dent, London Melbourne
Cott HB (1940) Adaptive coloration in animals. Methuen, London
Edmunds M (1974) Defence in animals. Longman, Harlow Essex
Eisner T (ed) (1960) Symposium 4 Chemical defensive mechanisms in arthropods. XI Internationaler Kongress für Entomologie, Wien. Verhandlungen III:245–293
Forsyth A, Miyata K (1984) Tropical nature. Charles Scribner's Sons, New York
Hinton HE (1955) Protective devices of endopterygote pupae. Trans Soc Brit Entomol 12:49–92
Hinton HE (1973) Natural deception. In: Gregory RL, Gombrich EH (eds) Illusion in nature and art. Gerald Duckworth, London
Pavan M (ed) (1960) Symposium 3. Insect chemistry. XI Internationaler Kongress für Entomologie, Wien, Verhandlungen III:1–243
Pavan M, Dazzini MV (1971) Toxicology and pharmacology – Arthropoda. Chem Zool 6:365–409

Richards OW, Davies RG (1977) Imms general text book of entomology. 10th ed. Chapman and Hall, London (2 vols)

Ward P (1979) Colour for survival. Orbis Publishing, London

Wickler W (1968) Mimicry in plants and animals. Weidenfeld and Nicolson, London; McGraw-Hill, New York Toronto (World University Library)

9 The Success of Terrestrial Arthropods

9.1 Criteria of Success

The Arthropoda and the vertebrates are usually regarded as the two most successful taxa so far to have evolved but, before we turn to consider the success of the terrestrial arthropods, the group with which this book is concerned, we should perhaps consider what is meant by successful. The marine crustacean *Calanus finmarchicus* has possibly the greatest biomass of any animal species, because it is adapted to planktonic life throughout the oceans of the world. It might, therefore, be regarded as more successful than *Thermocyclops schuurmanni,* which exists in the anoxic lower waters of a few small volcanic crater lakes in werstern Uganda — if we take biomass as the criterion of success. But is *Calanus* really more successful, or does it merely inhabit a larger environment?

The order Coleoptera comprises more species than all other animals put together. If, therefore, we use species diversity as the criterion of success, the Insecta is undoubtedly the most successful class of animals to have evolved. If we were to use other criteria, however, we might justifiably argue that mammals or birds should be awarded the accolade for success, or alternatively, the Nematoda. It has been said that, if all living organisms other than nematodes were to disappear, their outlines would still remain as ghostly images composed of the countless tiny nematode parasites that inhabit their tissues. Perhaps one should not attempt to be too precise. The number of species within a taxon is probably a better criterion than biomass, because it indicates the diversity of environments into which that taxon has radiated, whereas biomass merely reflects the absolute size of habitat to which a particular taxon has become adapted. It may, therefore, be justifiable to regard the terrestrial arthropods as being an eminently successful group. Let us now consider the reasons for this success.

9.2 Reasons for Success

The one factor common to all arthropods, and to which their success can undoubtedly be largely attributed, lies in the possession of a chitinous exoskeleton coupled with comparatively small size (Chap. 2.1), short life cycles, and genetic adaptibility, which have enabled them to colonize every conceivable terrestrial habitat. The insects are probably more successful than the Arachnida because the powers of flight enable them to migrate from one habitat to another, and exploit each to their own best advantage (Chap. 7.1). Yet the Coleoptera, the largest order of all, paradoxically displays a marked tendency towards a reduction of wings and even to flightlessness. Not all

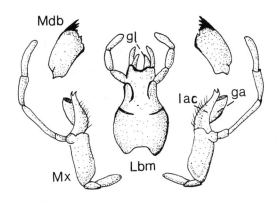

Fig. 75. Chewing mouthparts of a cockroach, *Mdb* mandibles; *Mx* maxillae; *Lbm* labium; *gl* glossa; *ga* galea; *lac* lacinia

insects are equally successful, either. There can be no doubt that the endopterygote orders Coleoptera, Diptera, Lepidoptera and Hymenoptera are pre-eminent. Although the Orthoptera and Isoptera are by no means unsuccessful, they do not exploit the diversity of habitats occupied by Endopterygota, which are able to utilize different modes of life when larvae and as adults.

9.3 Adaptability

On account of their short life-cycles, insects are able to adapt very rapidly to changing circumstances. The speed at which DDT-resistant strains of houseflies, mosquitoes and other insects evolved testifies to this. Arthropods have also been extremely successful in acquiring morphological adaptations of various kinds. Two examples only will be discussed in the following paragraphs, viz the acquisition of sucking mouth parts, and of ectoparasitic adaptations.

9.3.1 Evolution of Sucking Mouthparts

Sucking mouthparts have been evolved on a number of separate occasions from the primitive mandibulate chewing type as exemplified in cockroaches, grasshoppers and beetles (Fig. 75). Parallel evolution has occurred in different insect orders. For instance, in lice (Anoplura) the hypopharynx (Hyp) containing the salivary canal (sc), the maxillae (Mx) and the labium (Lbm) are in the form of long thin stylets which are retracted within the head when at rest, with their apices protected by the labrum (Lbr) (Fig. 76). In the Hemiptera the stylets are composed of mandibles, and maxillae (between which lies the salivary canal), supported by the labium (Fig. 77). In mosquitoes (Diptera, Nematocera) the labium again functions as a sheath and supports stylets consisting of mandibles, maxillae, hypopharynx (containing the salivary canal) and labrum (labrum-epipharynx) (Fig. 78). In higher Diptera, both mandibles and maxillae have disappeared and the mouthparts consist only of labium, labrum-epipharynx and hypopharynx (Fig. 79). The skin of the host is penetrated by the slender labium supported by enlarged maxillary palps. The mouthparts of fleas (Siphonaptera)

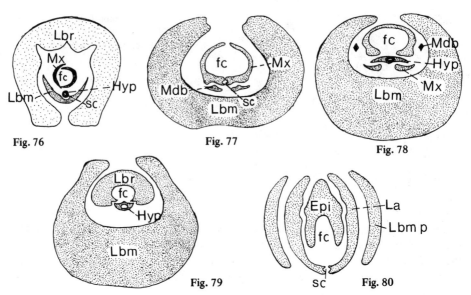

Fig. 76. Transverse section through the mouthparts of a louse (Anoplura). *Lbr* labrium; *Mx* maxillae; *Fc* food canal; *Hyp* hypopharynx; *Lbm* labium; *sc* salivary canal

Fig. 77. Transverse section through the mouthparts of a bug (Hemiptera). *Mx* maxilla; *Mdb* mandible; *Lbm* labium; *fc* food canal; *sc* salivary canal

Fig. 78. Transverse section through the mouthparts of a mosquito (Diptera Nematocera). *Lbr* labrum-epipharynx; *Mdb* mandible; *Hyp* hypopharynx (containing salivary canal); *Mx* maxilla; *Lbm* labium

Fig. 79. Transverse section through the mouthparts of a tsetse fly (Diptera Cyclorrhapha). *Lbr* labrum-epipharynx; *Hyp* hypopharynx; *Lbm* labium; *fc* food canal

Fig. 80. Transverse section through the mouthparts of a flea (Siphonaptera). *La* lacinia; *Lbm* p labial palp; *Epi* epipharynx; *fc* food canal; *sc* salivary canal

are quite different, and the homology of their components is not completely agreed upon. Mandibles are absent, and insertion is made by a pair of blade-like structures which are believed to have evolved from the laciniae (la) (Fig. 75) of the maxillae (Fig. 80). Between these is the epipharynx (Epi), enclosing the food canal. The epipharynx and laciniae are guided by the labial palps: the hypopharynx is rudimentary.

Some Hemiptera are predatory, some are bloodsuckers, while most feed upon plant juices. Lice, fleas, and many Diptera suck blood: Lepidoptera feed upon nectar. The coiled proboscis of butterflies and moths represents a modification of the galeae (ga) of the maxillae (Fig. 75). These are greatly elongated and held together by means of hooks and interlocking spines to form a tube connected with the mouth. Fluids are sucked in by the pumping of the pharynx. In its natural state, the proboscis is coiled, but it can be extended by its intrinsic musculature. In bees, the galeae and labial palps from a tube around the elongated glossae (gl) (Fig. 75), which are fused to form a tongue.

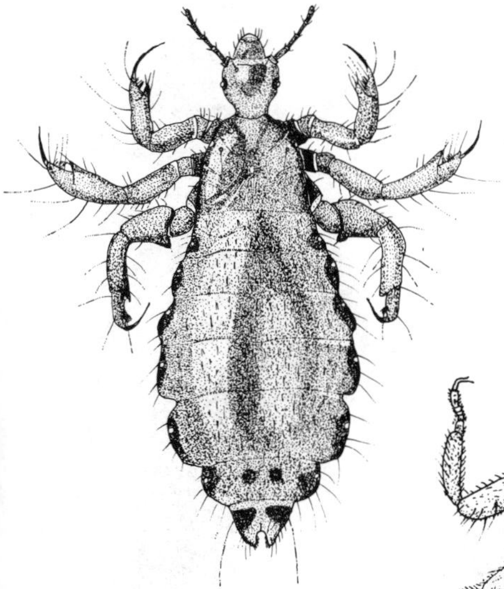

Fig. 81

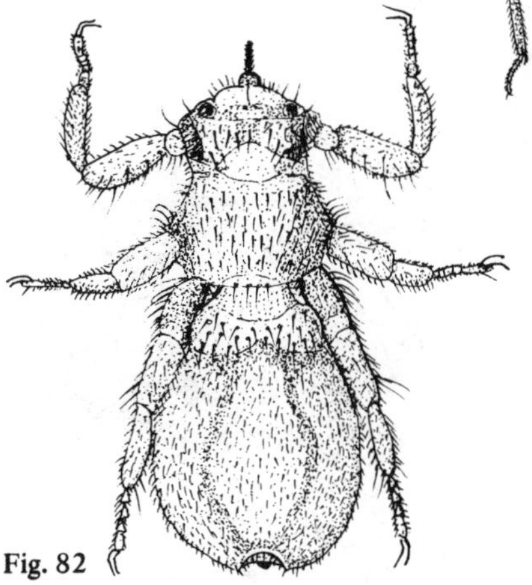

Fig. 82

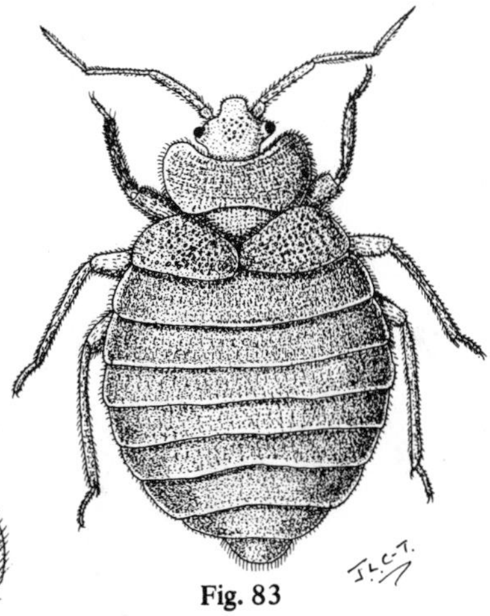

Fig. 83

Fig. 81. Human louse (*Pediculus*) (Siphunculata). (J.L. Cloudsley-Thompson)

Fig. 82. Louse-fly (*Melophagus*) (Diptera Hippoboscidae). (After Askew 1971)

Fig. 83. Bed-bug (*Cimex*) Hemiptera. (J.L. Cloudsley-Thompson)

9.3.2 Ectoparasitic Adaptations

Insects of several different orders have become adapted to an ectoparasitic mode of life. They tend to be flattened, wingless, with antennae reduced and directed backwards, so that they do not obstruct movement through the hair of the host, with reduced eyes — sense organs are less important to parasitic than to free-living forms — and a tough, rubbery, unsclerotized cuticle. They are not readily damaged when scratched by their hosts. The appearance of the louse (Fig. 81) and louse-fly (Fig. 82) are strikingly similar. (Some species of *Hippobosca* are fully winged, but others are wingless). The bed-bug (Fig. 83) and tick (Fig. 84) show similar adaptive characteristics, which appear even in *Arixenia* (Fig. 85), a large earwig, 2 cm long, which is to some extent parasitic on fruit bats in the Far East. The claws of different species of lice are particularly well adapted to the diameters of the hairs of the vertebrate host to which they cling, while bed-bugs, although wingless, lack the modified claws of permanent ectoparasites of birds and mammals.

Whereas all the parasitic arthropods so far mentioned are flattened dorso-ventrally, the flea (Fig. 86) is flattened laterally so that it is able rapidly to slip through the fur

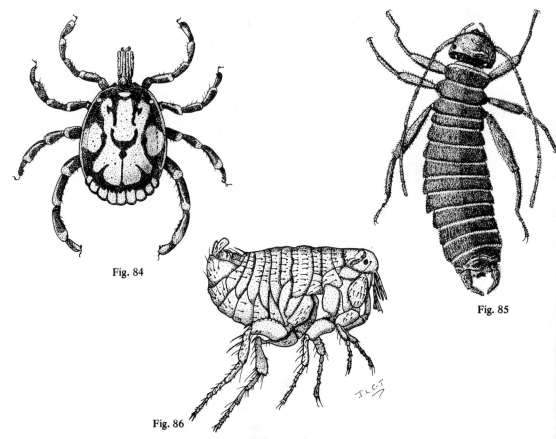

Fig. 84

Fig. 85

Fig. 86

Fig. 84. Tick (*Amblyomma*) (Acari). (After J. Smart)

Fig. 85. Parasitic earwig (*Arixenia esau*) (Dermaptera)

Fig. 86. Flea (*Xenopsylla*) (Aphaniptera). (J.L. Cloudsley-Thompson)

of the host. This also enables it to leap, because the back legs operate in a vertical plane. The combs or ctenidia on the thorax help to retain the flea amongst the fur or feathers of its host. They lock onto the hairs so that the flea is not dragged backwards by the scratching of the host. Whereas ticks and the dorso-ventrally compressed ectoparasitic insects are adapted to cling to the hairs of their hosts, like monkeys among the branches of trees, the flea is a swifter animal altogether and capable of moving rapidly through fur or feathers, like an antelope through tall grass. This analogy may seem fanciful, but it conveys an idea of the types of morphological adaptation of the different arthropods mentioned.

9.4 Conclusion

The two examples, of which brief synopses are given above, illustrate clearly what is meant by the adaptability of the terrestrial arthropods. As a taxon they are uniquely successful in terms of adaptability, diversity of structure, number of habitats occupied, numbers of individuals and of species, ecological dominance and so on, all associated with comparatively small multicellular bodies. (The most successful of larger animals, as we have seen, are the vertebrates). These same qualities make the terrestrial arthropods particularly interesting to study: and the proper justification for studying them lies in the fact that they are such fascinating animals.

Further Reading

Askew RR (1971) Parasitic insects. Heinemann Educational Books, London Edinburgh
Chapman RF (1969) The insects. Structure and function. English Universities Press, London
Hadley NF (1986) The arthropod cuticle. Sci Amer 254 (7):104–112
Richards OW, Davies RG (1977) Imms' general textbook of entomology, 10th edn (2 vols).
    Chapman and Hall, London; John Wiley, New York
Wigglesworth VB (1984) The principles of insect physiology, 8th edn. Chapman and Hall, London

Index

Only the more important references have been cited. Numerals in **boldface** type refer to illustrations.